The Morphology of Pteridophytes

Biological Sciences Editor

Professor A. J. Cain MA, D.PHIL
Professor of Zoology in the
University of Liverpool

The Morphology of Pteridophytes
The Structure of Ferns and Allied Plants

K. R. Sporne SCD, FLS

Fellow Emeritus of Downing College, Cambridge,
and University Lecturer in Botany

Hutchinson of London

Hutchinson & Co (Publishers) Ltd
3 Fitzroy Square, London W1

London Melbourne Sydney Auckland
Wellington Johannesburg and agencies
throughout the world

First published 1962
Second edition 1966
Reprinted 1968
Third edition 1970
Reprinted 1974
Fourth edition 1975
Reprinted 1979

Set in Monotype Times

Printed in Great Britain
by The Anchor Press Ltd and bound by
Wm Brendon & Son Ltd, both of Tiptree, Essex

ISBN 0 09 123860 9 (cased)
      0 09 123861 7 (paper)

Contents

To
H.H.T.
in grateful memory

Preface

For many years morphology was regarded as a basic discipline in the study of botany and, consequently, there have been many textbooks dealing with the subject. The pteridophytes have occupied varying proportions of these, and there have even been some textbooks devoted to a single group, e.g. the ferns, within the pteridophytes. Until the first edition of this book appeared in 1962, however, no book dealing solely with the pteridophytes had been published in the western hemisphere since 1936. Some of the old classics had been reprinted, but there was a need for a reappraisal of the old theories in the light of more recent knowledge.

Contrary to general belief, the study of morphology is a very live one and the rate at which advances are being made is accelerating. Exciting new fossils are being discovered and new techniques have been developed for investigating them. At the same time, new techniques are also being applied to the study of living organisms.

While the second and third editions took some account of discoveries that have been made since 1962, the time has now come for a major revision. A re-examination of Devonian fossils by new techniques, the discovery of many new genera of Devonian age and the amended dating of deposits hitherto thought to be Silurian, have led to a new understanding of early land-plant evolution. As a result, Chapter 2 has been almost completely re-written, several important genera originally described in that chapter are now dealt with in later ones, and there is an extra chapter describing the possible forerunners of seed-plants.

Most of the illustrations that appeared in the first edition have been used again, but they have been redistributed so as to be less crowded. Some, however, have been replaced by more recent interpretations, and there are many extra illustrations of the newly discovered fossil genera.

I should like to acknowledge again the help given, often unconsciously, by my colleagues in the Botany School, Cambridge, discussions with whom over the years have crystallized many of the ideas incorporated in this book. Most of all, I owe a debt of gratitude to my teacher and friend, the late Hugh Hamshaw Thomas, sc.d., f.r.s., who guided my first thoughts on the evolution of plants, who was a constant source of inspiration for more than twenty-five years and to whom this book is dedicated. It was he who first demonstrated to me that the study of living plants is inseparable from that of fossils, a fact which forms the basis for the arrangement of this book, in which living and fossil plants are given equal importance.

Finally, my grateful thanks are due to my wife for her helpful criticisms during the preparation of the manuscript, not only of the first edition, but also of subsequent editions.

K.R.S.
Cambridge, 1975

1 Introduction

The study of the morphology of living organisms is one of the oldest branches of science, for it has occupied the thoughts of man for at least 2500 years. Indeed, the very word 'morphology' comes from the ancient Greeks, while the names of Aristotle and Theophrastus occupy places of importance among the most famous plant morphologists. Strictly translated, morphology means no more than the study of form or structure. One may well ask, therefore, wherein lies the intense fascination that has captured the thoughts and imagination of so many generations of botanists from Aristotle's time to the present day; for the study of structure alone would be dull indeed. The answer is that, over the centuries, morphology has come to have wider implications, as Arber (1950) has explained in her *Natural Philosophy of Plant Form*. In this book she points out that the purpose of the morphologist is to 'connect into one coherent whole all that may be held to belong to the intrinsic nature of a living being'. This involves the study, not only of structures as such, but also of their relations to one another and their coordination throughout the life of the organism. Thus, morphology impinges on all other aspects of living organisms (physiology, biochemistry, genetics, ecology, etc.). Furthermore, the morphologist must see each living organism in its relationship to other living organisms (taxonomy) and to extinct plants (palaeobotany) whose remains are known from the fossil record of past ages extending back in time certainly 500 million years and probably as far back as 1000 million (some even say 2000 million) years. Clearly, the morphologist cannot afford to be a narrow specialist. He must be a biologist in the widest possible sense.

From taxonomy and palaeobotany, the plant morphologist is led naturally to the consideration of the course of evolution of plants (phylogeny), which to many botanists has the greatest fascination of all. However, it must be emphasized that here the morphologist is in the greatest danger of bringing discredit on his subject. His theories are not capable of verification by planned experiments and cannot, therefore, be proved right or wrong. At the best, they can be judged probable or improbable. Theories accepted fifty years ago may have to be abandoned as improbable today, now that more is known of the fossil record, and, likewise, theories that are acceptable today may have to be modified or abandoned tomorrow. It is essential, therefore, that the morphologist should avoid becoming dogmatic if he is ever to arrive at a true understanding of the course of evolution of living organisms.

Within the plant kingdom the range of size is enormous, for, on the one hand, there are unicellular algae and bacteria so small that individuals are visible only under the microscope, while, on the other hand, there are seed-bearing plants, such as the giant Redwoods of California and the Gums of Australia, some of which are probably the largest living organisms that the world has ever known. Accompanying this range of size, there is a corresponding range of complexity of internal anatomy and of life-history. Somewhere between the two extremes, both in structure and in life-cycle, come the group of plants known as pteridophytes, for they share with seed plants the possession of well-developed conducting tissues, xylem and phloem, but differ from them in lacking the seed habit. Internally, they are more complex than mosses and liverworts, yet in life-cycle they differ from them only in matters of degree.

The basic life-cycle, common to bryophytes and pteridophytes, is represented diagrammatically in Fig. 1. Under normal circumstances there is a regular alternation between a gametophyte (sexual) phase and a sporophyte (asexual) phase. The male gametes, produced in numbers from antheridia, are known as antherozoids, since they are flagellated and are able to swim in water, while the female gametes (egg cells) are non-motile and are borne singly in flask-shaped archegonia. Fusion between an egg cell and an antherozoid results in the formation of a zygote, which contains the combined nuclear material of the two gametes. Its nucleus contains twice as many chromosomes as either of the gamete nuclei and it is therefore des-

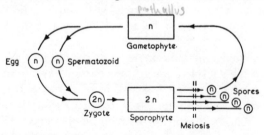

Fig. 1   Life-cycle of a homosporous pteridophyte

cribed as diploid. The zygote develops directly by mitotic divisions into the sporophyte which is, likewise, diploid. Ultimately, there are released from the sporophyte a number of non-motile spores, in the formation of which meiosis brings about a reduction of the nuclear content to the haploid number of chromosomes. The life-cycle is then completed when these spores germinate and grow, by mitotic divisions, into haploid gametophytes.

In mosses and liverworts, the dominant phase in the life-cycle is the gametophyte, for the sporophyte is retained upon it throughout its life and is either partially or completely dependent on it for nutrition. By contrast, among pteridophytes the sporophyte is the dominant generation, for it very soon becomes independent of the gametophyte (prothallus) and grows to a much greater size. Along with greater size is found a much greater degree of morphological and anatomical complexity, for the sporophyte is organized into stems, leaves and (except in the most ancient fossil pteridophytes and the most primitive living members of the group) roots. Only the sporophyte shows any appreciable development of conducting tissues (xylem and phloem), for although there are recorded instances of such tissues in gametophytes, they are rare and the amounts of xylem and phloem are scanty. Furthermore, the aerial parts of the sporophyte are enveloped in a cuticle in which there are stomata, giving access to complex aerating passages that penetrate between the photosynthetic palisade and mesophyll cells of the leaf.

All these anatomical complexities confer on the sporophyte the potentiality to exist under a much wider range of environmental conditions than the gametophyte. However, in many pteridophytes these potentialities cannot be realized, for the sporophyte is limited to those habitats in which the gametophyte can survive long enough for fertilization to take place. This is a severe limitation on those species whose gametophytes are thin plates of cells that lack a cuticle and are, therefore, susceptible to dehydration. Not all gametophytes, however, are limited in this way, for in some pteridophytes they are subterranean and in others they are retained within the resistant wall of the spore and are thus able to survive in a much wider range of habitats. It is notable that wherever the gametophyte is retained within the spore the spores are of different sizes (heterosporous), the larger megaspores giving rise to female prothalli which bear only archegonia, and the smaller microspores giving rise to male prothalli bearing only antheridia. Why this should be is not known with certainty, but two possible reasons come to mind, both of which probably operate together.

The first concerns the nutrition of the prothallus and the subsequent embryonic sporophyte. The retention of the gametophyte within a resistant spore wall severely limits its powers of photosynthesis and may even prevent it altogether. Hence, it is necessary for such a prothallus to be provided with abundant food reserves; the larger the spore, the more that can be stored within it. This may well account for the large size of the spores which are destined to contain an embryo sporophyte, but it does not explain why the prothalli should be unisexual (dioecious). This is most probably concerned with out-breeding. It is widely accepted that any plant which habitually under-goes inbreeding is less likely to produce new varieties than one which has developed some device favouring out-breeding, and that such a plant is at a disadvantage in a changing environment. It will tend to lag behind in evolution. Now, monoecious gametophytes (bearing both archegonia and antheridia) are much more likely to be self-fertilized than cross-fertilized, unless they are actually submerged in water. Yet, dioecious prothalli in a terrestrial environment would be at an even greater disadvantage, for they might never achieve fertilization at all, so long as the antherozoid has to bear the whole responsibility of finding the archegonium. This is where heterospory may operate to the advantage of plants with dioecious prothalli. Those spores which are destined to produce male prothalli need not carry large food reserves and can, therefore, afford to be reduced in size to the barest minimum. From the same initial resources, vast numbers of microspores can be produced and this will allow some of the responsibility for reaching an archegonium to be transferred to them. Blown by the wind, they can travel great distances and some, at least, will fall on a female prothallus in close proximity to an archegonium. Thus, when the male prothallus develops, the antherozoids liberated from the antheridia have only a short distance to swim and, in order to do so, need only a thin film of moisture. Under ordinary circumstances, the chances may be quite small that the particular microspore will have come from the same parent sporophyte as the megaspore and thus a fair degree of out-breeding will have been achieved. The relative emancipation from the aquatic environment provided by the heterosporous habit will confer on the sporophyte the freedom to grow almost anywhere that its own potentialities allow and the possibility of out-breeding will favour more rapid evolution of those potentialities. Most morphologists agree that the evolution of heterospory was a necessary step in the evolution of the seed habit and that, therefore, it is one of the most important advances in the whole story of land plant evolution.

The life-cycle of a typical heterosporous pteridophyte may be represented diagramatically as in Fig. 2.

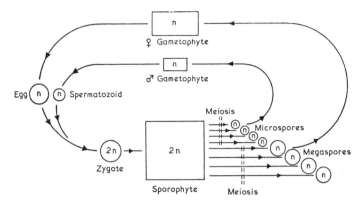

Fig. 2   Life-cycle of a heterosporous pteridophyte

The distinction between heterospory and homospory is one of the criteria used in the classification of pteridophytes, in accordance with the general belief that reproductive organs are a better guide to phylogenetic relationships than are vegetative organs. They are held to be more 'conservative', in being less susceptible to the immediate influence of the environment. Likewise, therefore, the manner in which the sporangia develop and the way in which they are borne on the sporophyte constitue important taxonomic characters.

The sporangium, in all pteridophytes, is initiated by the laying down of a cross-wall in a superficial cell, or group of cells. Since this wall is periclinal (i.e. parallel to the surface) each initial cell is divided into an outer and an inner daughter cell. If the sporogenous tissue is derived from the inner daughter cell, the sporangium is described as 'eusporangiate' and, if from the outer, as 'leptosporangiate'. This definition of the two types of sporangium is usually expanded to include a number of other differences. Thus, in leptosporangiate forms, the sporangium wall and the stalk, as well as the spores, are derived from the outer daughter cell, but, in eusporangiate forms, adjacent cells may become involved in the formation of part of the sporangium wall and the stalk (if any). Furthermore, the sporangium is large and massive in eusporangiate forms, the wall is several cells thick and the spore content is high, whereas, in leptosporangiate forms, the sporangium is small, the wall is only one cell thick and the spore content is low. Of these two types, the eusporangiate is primitive and the leptosporangiate advanced.

Until the early years of this century, it was widely believed that sporangia could be borne only on leaves and that such fertile leaves, known as 'sporophylls', were an essential part of all sporophytes. However, the discovery of Devonian pteridophytes that were completely without leaves of any kind, fertile or sterile, has led most morphologists to abandon this 'sporophyll theory'. It is now accepted that in some groups sporangia may be borne on stems, either associated or not with leaves, and in others actually on the leaves.

Important as reproductive organs are in classification, vegetative organs are nevertheless of considerable importance in classifying pteridophytes, for the shape, size, arrangement and venation of leaves (and even presence or absence of leaves) are fundamental criteria. It so happens that it is difficult, if not impossible, to devise a definition of the term 'leaf' that is entirely satisfying, but, for practical purposes, it may be said that among pteridophytes there are two very different types of leaf, known respectively as megaphylls and microphylls. The familiar fern frond is an example of the former; it is large, branches many times and has branching veins. By contrast, microphylls are relatively small, rarely branch and possess either a limited vascular supply or none at all; the leaf trace, if present, is single and either remains unbranched within the microphyll or, if it branches at all, it does so to a limited degree and in a dichotomous manner.

As might be expected, the leaf traces supplying microphylls cause little disturbance when they depart from the vascular system (stele) of the parent axis, whereas those supplying megaphylls are usually (though not invariably) associated with leaf gaps. A stele without leaf gaps is termed a protostele, the simplest type of all being the solid protostele. Fig. 3A illustrates its appearance diagrammatically as seen in transverse section. In the centre is a solid rod of xylem which is surrounded by phloem and then by pericycle, the whole stele being bounded on the outside by a continuous endodermis. Another variety of protostele is the medullated protostele, illustrated in Fig. 3B. In this the central region of the xylem is replaced by parenchyma. Yet other varieties of protostele will be described as they are encountered in subsequent chapters. Steles in which there are leaf gaps are known as dictyosteles, if the gaps occur frequently enough to overlap, and as solenosteles if they are more distantly spaced. Fig. 3C is a diagrammatic representation of a solenostele as seen in transverse section passing through a leaf gap. The most remarkable feature is the way in which the inside of the xylem cylinder is lined with phloem, pericycle and endodermis, as if these tissues had 'invaded' the

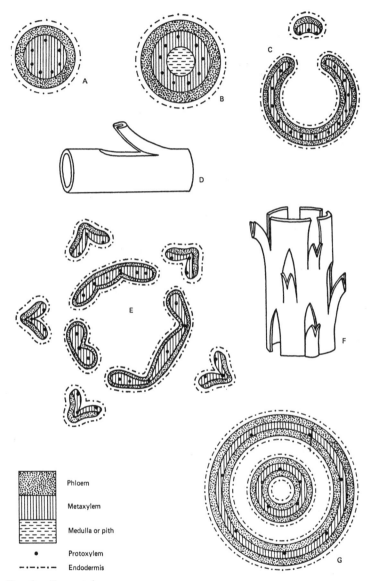

Fig. 3 Fern steles

A, solid protostele; B, medullated protostele; C, D, solenostele;
E, F, dictyostele; G, dicyclic stele.

central parenchymatous region (though, needless to say, the developmental processes do not involve any such invasion). Fig. 3E illustrates the structure of a dictyostele in which three leaf gaps are visible in the one transverse section. Frequently it happens that each leaf gap is associated with the departure of several leaf traces to the leaf, but in this example, for clarity, only one trace is shown supplying each leaf. The remaining portions of the stele are referred to as meristeles and, although in transverse section they appear to be unconnected, when dissected out and viewed as three-dimensional objects they are seen to form a network. Figs. 3D and 3F are perspective sketches of a solenostele and a dictyostele, respectively, from which the surrounding cortex and ground tissue have been removed in this way.

It must be pointed out at this stage that some morphologists use a different system of terminology and group together the medullated protostele and the solenostele as varieties of so-called siphonosteles, on the grounds that each has a hollow cylinder of xylem. The former they describe as an ectophloic siphonostele, because the phloem is restricted to the outside of the xylem, and the latter they describe as an amphiphloic siphonostele, because the phloem lies both outside and inside the xylem. This practice, however, has disadvantages. First, it tends to exaggerate the difference between the solenostele and the dictyostele – a difference that reflects little more than a difference in the direction of growth, for where leaves arise at distant intervals on a horizontal axis their leaf gaps are unlikely to overlap, whereas leaves on a vertical axis are often so crowded that their leaf gaps must overlap. Secondly, it overlooks the fundamental distinction between the solenostele and the medullated protostele – a physiological distinction depending on the position of the endodermis.

When gaps occur in a stele without any associated leaf traces, they are described as perforations and the stele is said to be perforated. Thus, there may be perforated solenosteles which, at a first glance, might be confused with dictyosteles; however, as soon as attention is paid to the relationship between leaf traces and perforations, the distinction becomes clear. When more than one stele is visible in any one transverse section the plant is described as polystelic. Yet another variant is the polycyclic stele, in which there are two or more co-axial cylinders of conducting tissue (Fig. 3G).

All the vascular systems mentioned so far are composed entirely of primary tissues, i.e. tissues formed by the maturation of cells laid down by the main growing point (apical meristem). It is customary to draw a rough distinction between tissues that differentiate before cell elongation has finished and those that differentiate only after such growth has ceased. In the former case, the xylem and phloem are described as protoxylem and protophloem. They are so constructed that they can still alter their shape and can, thereby, accommodate to the continuing elongation of the adjacent cells. Accordingly, it is usual for the lignification of protoxylem elements to be laid down in the form of a spiral, or else in rings. Metaxylem and metaphloem elements, by contrast, do not alter their size or shape after differentiation.

The order in which successive metaxylem elements mature may be centripetal or centrifugal. When the first xylem to differentiate is on the outside and differentiation proceeds progressively towards the centre, the xylem is described as exarch and all the metaxylem is centripetal. When the protoxylem is on the inner side of the metaxylem and differentiation occurs successively away from the centre, the xylem is described as endarch and all the metaxylem is centrifugal. A third arrangment is known as mesarch, where the protoxylem is neither external nor central and differentiation proceeds both centripetally and centrifugally. In Figs. 3A–C the xylem is mesarch, while in Fig. 3G it is endarch.

In addition to primary vascular tissues, some pteridophytes possess a vascular cambium from which secondary xylem and secondary phloem are formed. Cambial cells possess the power of cell division even though the surrounding tissues may have lost it; they may either have retained this power throughout the lapse of time since they were laid down in the apical meristem, or they may have regained it after a period of temporary differentiation. While relatively uncommon in living pteridophytes, a vascular cambium was widely present in coal-age times, when many members of the group grew to the dimensions of trees. Just as, at the present day, all trees develop bark on the outside of the trunk and branches by the activity of a cork cambium, so also did these fossil pteridophytes. In some, the activity of this meristem was such that the main bulk of the trunk was made up of the periderm which it produced.

Any attempt to interpret modern pteridophytes must clearly take into account their forerunners, now extinct, in the fossil record. This involves some understanding of the ways in which fossils came to be formed and of the extent to which they may be expected to provide information useful to the morphologist. A fossil may be defined as 'anything which gives evidence that an organism once lived'. Such a wide definition is necessary to allow the inclusion of casts, which are

fossil

formation

no more than impressions left in the sand by some organism. Yet, despite the fact that casts exhibit nothing of the original tissues of the organisms, they are nevertheless valuable in showing their shape. At the other extreme are petrifactions, in which the tissues are so well preserved by mineral substances that almost every detail of the cell walls is visible under the microscope. Between these two extremes are fossils in which decay had proceeded, to a greater or lesser degree, before their structure became permanent in the rocks.

Under certain anaerobic conditions (e.g. in bog peat and marine muds), and in the absence of any petrifying mineral, plant tissues slowly turn to coal, in which little structure can be discerned, apart from the cuticles of leaves and spores. Portions of plants that are well separated from each other by sand or mud during deposition give rise to fossils known as mummifications or compressions. From these, it is often possible to make preparations of the cuticle, by oxidizing away the coally substance with perchloric acid. Examination under a microscope may then reveal the outlines of the epidermal cells, stomata, hairs, papillae, etc. In this way, a great deal can be discovered from mummified leaves. Mummified stems and other plant organs, however, yield less useful results. Even their shape needs careful interpretation, because of distortion during compression under the weight of overlying rocks.

By far the most useful fossils to the palaeobotanist are those in which decay was prevented from starting, by the infiltration of some toxic substance, followed by petrifaction before any distortion of shape could occur. Such are, unfortunately, rare indeed. The most beautiful petrifactions are those in silica, but carbonates of calcium and magnesium are also important petrifying substances. Iron pyrites, while common, is less satisfactory because the fine structure of the plant is more difficult to observe. While it has often been said that during petrifaction the tissues are replaced molecule by molecule, this cannot be correct, for the 'cell walls' in such a fossil dissolve less rapidly in etching fluids than does the surrounding matrix. This fact forms the basis of a rapid technique for making thin sections of the plant material (Joy, Willis and Lacey, 1956). A polished surface is etched for a brief period in the appropriate acid and the cell walls that remain projecting above the surface are then embedded in a film of cellulose acetate. This is stripped off and examined under the microscope without further treatment, the whole process having taken no longer than ten minutes.

While it is frequently possible to discern the type of thickening on the walls of xylem elements, it is, however, rarely possible to make out much detail in the phloem of fossil plants, for this is the region which decays most rapidly. Furthermore, most fossils consist only of fragments of plants. It is then the task of the palaeobotanist to reconstruct, as best he can, from such partly decayed bits, the form, structure and mode of life of the whole plant from which they came. There is small wonder, then, that this has been achieved for very few fossil plants. Many years may elapse before it can be said with any certainty that a particular kind of leaf belonged to a particular kind of stem and, in the meantime, each must be described under a separate generic and specific name. In this way, the palaeobotanist becomes unavoidably encumbered by a multiplicity of such names.

For convenience of reference, the history of the Earth is divided into four great eras. The first of these, the pre-Cambrian era, ended about 500 million years ago and is characterized by the scarcity of fossils, either of animals or of plants. Then came the Palaeozoic era, characterized by marine invertebrates, fishes and amphibians, the Mesozoic by reptiles and ammonites and, finally, the Cainozoic, extending to the present day, characterized by land mammals. These major eras are again divided into periods (systems) and then subdivided again, chiefly on the basis of the fossil animals contained in their strata. While such a scheme is clearly satisfactory to the zoologist, it is less so to the botanist, for the plants at the beginning of one period (e.g. the Lower Carboniferous) are less like those of the end of the period (the Upper Carboniferous) than like those of the end of the previous period (the Upper Devonian). Thus, it is more usual for the palaeobotanist to speak of the plants of the Upper Devonian/Lower Carboniferous than of the plants of the Carboniferous period.

The sequence of the various geological periods is summarized as a table (p. 20), in which the time scale is based on information from Harland, Smith and Wilcock (1964). Brief notes are included to indicate the kind of vegetation that is believed to have existed during each period, but a word of caution is necessary on this matter. It must always be remembered that our knowledge of past vegetation is based entirely on those fragments of plants that happened to become fossilized and which, furthermore, happen to have been unearthed. Such fragments are clearly no more than a minute sample of the world's vegetation. They are not even a random sample, for there must have been a bias in favour of those plants which were growing in places where fossilization was likely to occur.

The time of appearance of the first pteridophytes has been the subject of debate for many years, but it is now generally accepted

Geological periods in the northern hemisphere

Era	Period	Age in 10^6 years	Type of vegetation
	Quaternary	1	Modern
Cainozoic	Upper Tertiary, Pliocene	11	Modern
	Miocene	25	
————	Lower Tertiary, Oligocene	40	Modern with tropical
	Eocene	60	plants in Europe
	Upper Cretaceous	100	
Mesozoic	Lower Cretaceous	136	Gymnosperms dominant (Conifers and
	Upper Jurassic	162	Bennettitales)
	Lower Jurassic (Liassic)	190	Luxuriant forests of Gymnosperms and
	Upper Triassic (Rhaetic)	205	Ferns
————	Lower Triassic (Bunter)	225	Sparse desert flora with Gymnosperms (Conifers and
	Upper Permian	240	Bennettitales)
	Lower Permian	280	Tall swamp forests with early Gymnosperms,
	Upper Carboniferous (Coal Measures)	325	Tree-Lycopods, Calamites and Ferns
	Lower Carboniferous	345	Early Gymnosperms, large Tree-Lycopods
	Upper Devonian	359	and Ferns
Palaeozoic	Middle Devonian	370	Herbaceous Lycopods and early Ferns
	Lower Devonian	395	Herbaceous marsh plants (*Cooksonia* and
	Upper Silurian		*Zosterophyllum*)
	Silurian	430	Marine algae
	Ordovician	500	Marine algae
	Cambrian	570	Marine algae, but some evidence of land plants, too
Pre-Cambrian		4500?	Fungi and Bacteria reported to have occurred 2000 million years ago

that this event must have occurred during late Silurian times. Claims that cutinized spores prove that land plants existed as early as the Cambrian are now largely discounted (Chaloner, 1967). Not only do pteridophytes possess cutinized spores and a waxy cuticle with stomata, but they also have a vascular system containing lignified xylem elements, and this combination of features does not appear in the fossil record until close to the boundary between the Silurian and the Devonian periods, some 400 million years ago.

We turn now to the classification of pteridophytes. The first object of any classification must be to group together similar organisms and to separate dissimilar ones. In the process the group is subdivided into smaller groups, each defined so as to encompass the organisms within it. In the early days of taxonomy, when few fossils were known, these definitions were based on living plants. Then, as more and more fossils were discovered, modifications became necessary in order to accommodate them, and a number of problems arose. The first arises from the fact that a fossil plant, even when properly reconstructed, is known only at the stage in its life-cycle when it died. Other stages in its life-cycle, or in its development, may never be discovered. Yet, the classification of living organisms may (and indeed should) be based on all stages of the life-cycle. The second problem concerns the difficulty, when new fossils are discovered, of deciding whether to modify the existing definitions of groups or whether to create new groups. Too many groups would be liable to obscure the underlying scheme of the classification and too few might result in each group being so wide in definition as to be useless. The scheme on which this book is based is a modification of the one used by Reimers in the 1954 edition of Engler's *Syllabus der Pflanzenfamilien* (Melchior and Werdermann, 1954); it has been chosen because it seems to strike a balance in the number and the size of the groups that it contains. (An asterisk is used throughout to indicate fossil groups.)

PTERIDOPHYTES

A PSILOPSIDA*
1 Rhyniales*
2 Trimerophytales*
3 Zosterophyllales*

B PSILOTOPSIDA
Psilotales

C LYCOPSIDA
1 Protolepidodendrales*
2 Lycopodiales
3 Lepidodendrales*
4 Isoetales
5 Selaginellales

D SPHENOPSIDA
1 Hyeniales*
2 Sphenophyllales*
3 Calamitales*
4 Equisetales

E PTEROPSIDA
a Primofilices*
1 Cladoxylales*
2 Coenopteridales*

b Eusporangiatae
1 Marattiales
2 Ophioglossales

c Osmundidae
Osmundales

d Leptosporangiatae
1 Filicales
2 Marsileales
3 Salviniales

F PROGYMNOSPERMOPSIDA*
1 Aneurophytales*
2 Protopityales*
3 Archaeopteridales*

2 Psilopsida

Extinct plants. Only the sporophytes known. Rootless, with rhizomes and more or less dichotomous aerial axes, either naked or bearing small lateral appendages. Protostelic. Sporangia thick-walled, homosporous, terminal or lateral.

1 Rhyniales*
    Rhyniaceae*   *Rhynia*, Horneophyton** (=*Hornea*), *Yarravia**
    Cooksoniaceae*   *Cooksonia**
2 Trimerophytales*   *Trimerophyton*, Psilophyton*, Pertica**
3 Zosterophyllales*
    Zosterophyllaceae*   *Zosterophyllum*, Rebuchia**
    Gosslingiaceae*   *Gosslingia*, Crenaticaulis*, Sawdonia*,*
                        Kaulangiophyton*

The first member of this group ever to be described was *Psilophyton princeps* (Dawson, 1859), but for many years little notice was taken of this discovery. Indeed, so different was it from any preconceived ideas of early land plants that many botanists regarded it as a figment of the imagination. However, when Kidston and Lang (1917, 1920a and 1921a) described a number of similar plants from the Devonian rocks of Rhynie, in Scotland, the idea at last became accepted that plants with such a simple organization really had existed.

RHYNIALES

The chert deposits at Rhynie, originally thought to be of Middle Devonian age, are now generally believed to be late Lower Devonian (Emsian). They are about 2·5 metres thick and are thought to represent a peat bog which became infiltrated with silica. In this way the plant remains became preserved, some of them with great perfection. The chief plants to have been described from these deposits are *Rhynia major*, *Rhynia Gwynne-Vaughani*, *Horneophyton Lignieri* and *Asteroxylon Mackiei*. Of these, the first three had naked, forking axes, and are now grouped together in the Rhyniales, along with

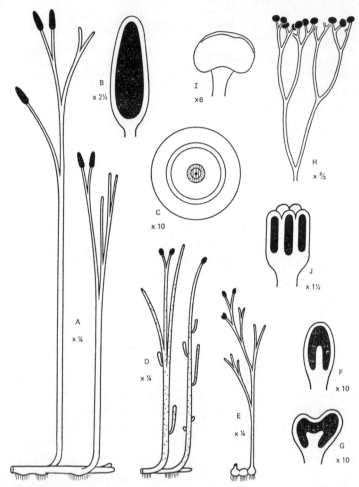

Fig. 4   Rhyniales

A–C, *Rhynia major*: A, reconstruction; B, sporangium; C, t.s. stem.
D, *Rhynia Gwynne-Vaughani*, reconstruction.
E–G, *Horneophyton Lignieri*: E, reconstruction; F, G, sporangia.
H–I, *Cooksonia caledonica*: H, reconstruction; I, sporangium, showing
dehiscence line; J, *Yarravia oblonga*, sporangia.

(A, D, E, after, and B, F, G, based on Kidston and Lang; H, I, after
Edwards; J, based on Lang and Cookson)

Yarravia from Lower Devonian rocks of Australia, and *Cooksonia* from Upper Silurian and Lower Devonian rocks of the British Isles, Czechoslovakia and the USA.

The general appearance of *Rhynia major* is illustrated in Fig. 4A. It had a horizontal rhizome which branched in a dichotomous manner and bore groups of unicellular rhizoids at intervals. The tips of some rhizomes turned upwards and grew into aerial stems as much as 50 cm high and up to 6 mm in diameter. These also branched dichotomously and some of them terminated in pear-shaped sporangia up to 12 mm long. The aerial parts were smooth and covered with a cuticle in which stomata were sparingly present, their presence indicating that the stems were green and photosynthetic. In transverse section (Fig. 4C) the stems are seen to have had a cortex differentiated into two regions, often separated by a narrow zone of cells with dark contents. Whereas the outer cortex was of densely packed cells, the inner cortex had abundant inter-cellular spaces with direct access to the stomata; for this reason the inner cortex is presumed to have been the main photosynthetic region. In the centre was a solid rod of xylem tracheids with annular thickenings. The smallest tracheids, believed to represent the protoxylem, occupied a central position and therefore the stele is described as having been 'centrarch'. Around the xylem was a zone of thin-walled cells which have recently been shown to have had sieve-like areas in the lateral walls (Satterthwaite and Schopf, 1972). Earlier suggestions that this zone represented one of phloem sieve cells now, therefore, receives strong support. The sporangium (Fig. 4B) had a massive wall, about five cells thick, apparently without any specialized dehiscence mechanism, and within it were large numbers of spores about 65μ in diameter. The fact that these spores were arranged in tetrads is taken to prove that they were formed by meiosis and that the plant bearing them represented the sporophyte generation.

Rhynia Gwynne-Vaughani (Fig. 4D) was a smaller plant than *R. major*, attaining a height of only 20 cm. It was similar in having a creeping dichotomous rhizome with groups of rhizoids, but the aerial parts of the plant differed in several respects; small hemispherical lumps were scattered over the surface and, besides branching dichotomously, the plant was able to branch adventitiously. An interesting feature of the adventitious branches was that the stele was not continuous with that of the main axis. It is possible that they were capable of growing into new plants if detached from the parent axis, thereby providing a means of vegetative propagation. The sporangia were only 3 mm long and the spores, too, were smaller than those of

R. *major*. In other respects (internal anatomy, cuticle, stomata, etc.) the two species were very similar indeed.

What the gametophytes of *Rhynia* looked like is not known with certainty, although the discovery of living gametophytes of *Psilotum* containing vascular tissue encouraged suggestions that some of the axes which have been described as rhizomes could have been gametophytes (Merker, 1958 and 1959; Pant, 1962; Lemoigne, 1968). Lemoigne (1970) goes so far as to suggest that only one species of *Rhynia* existed in the Scottish locality and that all the axes hitherto identified as *R. Gwynne-Vaughani* were gametophytes, while those identified as *R. major* were sporophytes. The small lumps on the surface of the former are regarded by him as protruding archegonia or antheridia. However, although Lemoigne has published illustrations of structures which he believes to have been archegonia, they are not completely convincing.

Horneophyton Lignieri (Fig. 4E) was smaller even than *Rhynia Gwynne-Vaughani*, its aerial axes being only some 13 cm high and only 2 mm in maximum diameter. It was first described under the generic name *Hornea*, but in 1938 it was pointed out that this name had already been used for another plant and a new name was proposed, *Horneophyton*. The aerial axes were like those of *Rhynia major*, in being quite smooth and in branching dichotomously without any adventitious branches. In its underground organs, *Horneophyton* was very different, for it had short lobed tuberous corm-like structures. From their upper side aerial axes grew vertically upwards and on their lower side were unicellular rhizoids. The stele of the aerial axis did not continue into the tuber, which was parenchymatous throughout. Most of the tubers contained abundant non-septate fungal hyphae, whose mode of life has been the subject of some speculation. By analogy with other groups of pteridophytes, it has been suggested that there was a mycorrhizal association but, as Kidston and Lang (1921b) pointed out, some well-preserved tubers showed no trace whatever of fungus. This fact suggests that, instead of being mycorrhizal, the fungus was a saprophyte which invaded the tissues of the tuber after death.

Another feature of interest, peculiar to *Horneophyton*, was the presence of a sterile columella in the sporangium (Fig. 4F), a feature reminiscent of the mosses. Kidston and Lang described one example of a bifid sporangium, as illustrated in Fig. 4G. More recently, El-Saadawy (1966) has shown that the sporangium sometimes forked repeatedly, with up to three successive dichotomies, and Eggert (1974) comes to substantially the same conclusion. It is interesting that, in such branching sporangia, the columella was also branched. This leads one to suppose that the stem apex could be transformed into a sporangium at any stage, even during the process of dichotomizing, and rules out any idea of the sporangium being borne by a special organ to which the name 'sporangiophore' might be given.

The Australian genus *Yarravia* (Fig. 4J) has been interpreted as a slender unbranched axis, terminating in a radially symmetrical group of five or six sporangia, partly fused into a synangium, about 1 cm long. Although Lang and Cookson (1935) who first described this genus, were unable to demonstrate the presence of spores within the sporangia this interpretation is widely accepted and has been used as the starting point for phylogenetic speculations as to the nature of the pollen-bearing organs of fossil seed-plants, and even of their seeds. We have no idea what kind of plant bore these peculiar reproductive bodies, for they had become detached before fossilization took place.

Similar remarks apply to *Cooksonia*, whose naked forking axes with terminal sporangia represent merely the terminal portions of a plant the lower regions of which are unknown. Five species have been described, of which the largest specimens so far discovered are no more than 7 cm long and 1·5 mm wide. The Cooksoniaceae differ from the Rhyniaceae in having relatively short and wide sporangia, varying in shape from reniform through spherical to oval. Fig. 4H illustrates a reconstruction of a recently described species from the early Lower Devonian (Dittonian = Gedinnian?) Old Red Sandstone of Scotland (Edwards, 1970a) while Fig. 4I illustrates the way in which its reniform sporangium dehisced.

Some specimens of *Cooksonia*, e.g. *C. pertonii*, from the Downtonian of Wales, are sufficiently well preserved for a thin vascular strand of annular tracheids to be demonstrated within the delicate axes, and for spores with tri-radiate scars to be extracted from the sporangia. There can be little doubt, therefore, that these were land plants and, now that Downtonian strata are generally regarded as Upper Silurian, *Cooksonia* must be accepted as the earliest vascular plant so far discovered. For this reason, its mode of branching and the way in which its sporangia were borne take on a special interest. In Lower Devonian rocks there are remains of vascular plants exhibiting four main types of organization viz. Rhyniales, Zosterophyllales and Trimerophytales (among Psilopsida) and Protolepidodendrales (among Lycopsida). The question arises as to whether *Cooksonia* could have provided the starting point for the evolution of all other vascular plants, or whether several different and unrelated types of vascular plant were emerging from aquatic environments at about the same time.

TRIMEROPHYTALES

Very little stretching of the imagination is needed to derive members of the Trimerophytales from plants like *Cooksonia*. Their sporangia were borne terminally on dichotomous axes with a centrarch rod of xylem tracheids. The most obvious difference was the tendency for successive unequal dichotomies to produce a monopodial system with a main axis bearing lateral branch systems of limited growth, as illustrated in Fig. 5. Another difference lay in the mode of dehiscence of their sporangia which was by means of a longitudinal slit. While most members of the group had smooth axes, some had spine-like or glandular outgrowths ('enations'). This fact has led to great confusion as to the correct diagnosis of the genus *Psilophyton*, a confusion which was apparent in earlier editions of this book, as in other textbooks of palaeobotany, and which has only recently been cleared up (Hueber and Banks, 1967; Hueber, 1971).

Psilophyton princeps (Fig. 5C) was first described by Dawson (1859) from late Lower Devonian (Emsian) rocks of the Gaspé Peninsula, Quebec. Early reconstructions of the plant showed a system of horizontal rhizomes, but it is now believed that these probably belonged to a different plant altogether. As things now stand, it must be confessed that we know nothing of the lower parts of the plant and that we cannot be certain, therefore, how tall it was. Its main axes were up to 1 cm in diameter and at least 1 m tall. The xylem strand was massive, compared with that of the Rhyniales and was made up of scalariform tracheids. The abundant enations clothing the axes have been variously described as leaves, spines and thorns, but their tips appear to have been glandular, they lacked stomata and they lacked vascular supply, so none of the descriptions seems to be really appropriate. Since stomata were present in the cuticle covering the

---

Fig. 5   Trimerophytales

A, *Psilophyton Forbesii*; B, *Pertica quadrifaria*; C, *Psilophyton princeps*; D, *Psilophyton Dawsonii*; E, *Trimerophyton robustius*.
(All are reconstructions of fertile shoots, with sporangia shown in black.)

(A, after Andrews, Kasper and Mencher; B, based on Kasper and Andrews; C, D, after Hueber; E, based on Hopping.)

A
x ⅛

B
x 1

C
x ½

D
x ½

E
x 2

Fig. 5

axis, it is presumed that the principal site of photosynthesis was the cortex of the axis itself.

Psilophyton Dawsonii, from the Lower Devonian of Ontario, was completely without enations (Fig. 5D), as was *P. Forbesii*, described by Andrews, Kasper and Mencher (1968) from Maine, USA, (Fig. 5A). *Trimerophyton robustius* (Fig. 5E) from the Gaspé Peninsula was described originally as a species of *Psilophyton*, but it differed markedly in its mode of branching. The lateral branches, arranged spirally on the main axis, trifurcated near their points of origin. Each of these second order branches then divided unequally to give third order branches which dichotomized twice into fertile pedicels bearing terminal clusters of sporangia. The sporangia were large and robust, up to 5 mm long and 1·5 mm wide (Hopping, 1956).

Pertica quadrifaria (Fig. 5B) has recently been described by Kasper and Andrews (1972) from middle Lower Devonian (Siegenian) deposits in northern Maine, USA. Growing perhaps to a height of 1 m, this plant had a main axis up to 1·5 cm in diameter, on which sterile and fertile lateral branch systems were arranged in four vertical rows. The sporangia, borne in dense terminal clusters, were up to 3 mm long and 1 mm wide.

Particular interest attaches to the Trimerophytales because as Banks (1968) points out, they foreshadow the still more complex branching patterns of the Cladoxylales, Coenopteridales and Progymnospermopsida, towards which he believes they evolved.

ZOSTEROPHYLLALES

Whether the Zosterophyllales represent an off-shoot of the same evolutionary line is open to considerable doubt, for they differed not only in having lateral instead of terminal sporangia, but also in having exarch instead of centrarch xylem strands (at least, in all those genera whose internal anatomy is known). They shared these two

---

Fig. 6   Zosterophyllales

A–C, *Zosterophyllum*: A, reconstruction of *Z. myretonianum*; B, reconstruction of *Z. rhenanum*; C. sporangial region of *Z. fertile*; D, E, *Kaulangiophyton akantha*: D, reconstruction; E, sporangial region; F, *Sawdonia ornata*, sporangial region. G, *Gosslingia breconensis*, reconstruction.

(A, after Walton; B, Kräusel and Weyland; C, Leclercq; D, E, Gensel, Kasper and Andrews; F, Hueber; G. Edwards.)

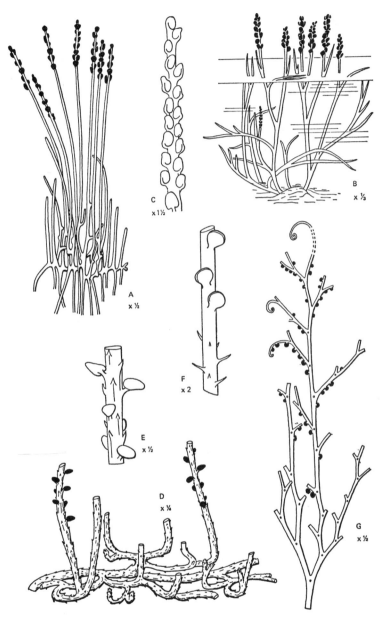

A
x ½

C
x 1½

B
x ⅓

F
x 2

E
x ½

D
x ¼

G
x ½

Fig. 6

features with the Lycopsida, to which they could conceivably be related, but it is difficult to see how they could have evolved from the Rhyniales within the period of only a few million years that elapsed between the first appearance of *Cooksonia*, in the uppermost Silurian, and of *Zosterophyllum*, in the lowermost Lower Devonian (Gedinnian).

Although no one species of *Zosterophyllum* was worldwide in its distribution, the genus was wide spread, being represented by *Z. rhenanum* in Germany, *Z. fertile* in Belgium, *Z. australianum* in Australia, *Z. minor* in Russia and *Z. myretonianum* in Russia and Scotland (Edwards, 1973).

Fig. 6A illustrates a reconstruction of *Z. myretonianum*, in which the lowermost parts of the plant underwent H-branching, thereby producing a tufted growth habit. The erect branches 15 cm or more tall and 2 mm in diameter, were smooth, sparsely forking and provided with stomata in the upper regions (Walton, 1964). From this it was concluded that at least the upper regions were aerial, even though the lower parts might have been under water or buried in mud. Fig. 6B is of a reconstruction of *Z. rhenanum* (Kräusel and Weyland, 1935) which has been widely reproduced in textbooks and which suggests that the plant was mostly submerged, only the reproductive organs rising above the surface of the water. This conclusion, based on the flattened nature of the axes, is not now thought to be justified, and the axes are thought to have become flattened during preservation.

The sporangia of *Zosterophyllum* were grouped together into terminal spikes, as illustrated in Figs. 6A–C. In some species, the sporangia were arranged spirally but in others (sometimes classed together in the subgenus *Platyzosterophyllum*) they were in two opposite rows, as in *Z. fertile* (Fig. 6C). The sporangia were ovoid or reniform, and dehisced by means of a slit running across the distal face of the sporangium. The xylem strand in *Z. llanoveranum* was elliptical in the fertile region, where the sporangia were in two rows, but there are indications that it was cylindrical lower down (Edwards, 1969).

Rebuchia is a new name for *Bucheria*, which some palaeobotanists use as a 'form genus' for detached spikes of sporangia looking like those of *Zosterophyllum*, but which Hueber (1972) regards as sufficiently distinct from *Zosterophyllum* to warrant separate generic status. Not only were the sporangia attached in two opposite rows, but they were also bent round in such a way as to lie on one side of the spike. The spikes were borne on dichotomously branching axes, the vegetative branches of which gradually tapered to blunt points.

All the other genera belonging to the Zosterophyllales bore their sporangia in a more or less random manner along the axes of the plant, as in *Gosslingia* (Edwards, 1970b), illustrated in Fig. 6G. First described by Heard (1927) from the Old Red Sandstone of the Brecon Beacons area of South Wales, *Gosslingia* is thought to have grown to a height of some 50 cm (although the largest portion so far found was only 15 cm long). Branching was by unequal dichotomy, which gives an impression of a monopodial system with a main axis and smaller alternate laterals. The maximum thickness of the axes was 4 mm, while the smallest axes, whose tips were circinately coiled, were only 0·5 mm across.

The xylem strand was elliptical in transverse section, was made up of tracheids with scalariform or occasionally reticulate pitting, and had exarch protoxylem. Just beneath each dichotomy a minute tubercle is visible, but what this represented in the living plant is debatable. The vascular strand trifurcated at each dichotomy of the axis, and one of the three strands extended into the tubercle. The fact that all the tubercles occurred on the same side has not been properly explained, but it is more in keeping with the concept of a plant with an upper and a lower side than of one with a right and a left side. There have been suggestions that each tubercle was a branch primordium, or bud, but against this interpretation is the fact that the vascular bundle running towards the tubercle was actually larger than either of the branch bundles. Edwards suggests that it is more likely to represent the base of a branch which abscinded before preservation, or was lost during preservation.

Some support for this interpretation is provided by the discovery of *Crenaticaulis*, in which a short axis, 7 mm long, has actually been found growing out from a similar axillary position (Banks and Davis, 1969). *Crenaticaulis* was very similar to *Gosslingia* in many ways: in its branching habit; in the lateral position of its sporangia; in its internal vascular anatomy; and in the fact that its axillary tubercles were on one side of the stem only. The main difference was in the tooth-like emergences on the surface of the stem of *Crenaticaulis*. Banks and Davis liken both the axillary tubercles of *Gosslingia* and the short outgrowths of *Crenaticaulis* to the 'rhizophores' that develop from angle meristems in some species of *Selaginella*.

Fig. 6F illustrates a small portion of a fertile axis of *Sawdonia*. The glandular spine-like enations closely resemble those of *Psilophyton princeps*, with the result that, when lateral sporangia were discovered,

B

the definition of the genus *Psilophyton* was thrown into a state of confusion lasting for several years. However, the internal anatomy of *Sawdonia* is like that of the Zosterophyllales, in being elliptical in cross-section and in being exarch.

Kaulangiophyton, from the uppermost Lower Devonian or the lowermost Middle Devonian of Maine, USA, and illustrated in Figs. 6D and E, is an enigmatic genus. It had axes up to 9 mm in diameter, branching by very wide dichotomies (H- or K-patterns), some remaining horizontal and others becoming erect. The axes bore short stout spines and ovoid sporangia on short lateral stalks. Nothing is known of its internal anatomy, but Gensel, Kasper and Andrews (1969) suggest that its affinities are nearer to the Zosterophyllales than to any other group. However, they emphasize that the discovery of *Kaulangiophyton* diminishes the sharpness of the distinction between the Psilopsida and the Lycopsida. The idea of an evolutionary origin of the Lycopsida from the Zosterophyllales, therefore, now seems more plausible than it did before the discovery of this interesting plant.

3 Psilotopsida

Sporophyte rootless, with dichotomous rhizomes and aerial branches. Lateral appendages spirally arranged, scale-like or leaf-like. Protostelic (either solid or medullated). Sporangia thick walled, homosporous, terminating very short lateral branches. Antherozoids multiflagellate.

Psilotales
  Psilotaceae *Psilotum*
  Tmesipteridaceae *Tmesipteris*

This small group of plants is one of great interest to morphologists because its representatives are at a stage of organization scarcely higher than that of some of the earliest land plants, despite the fact that they are living today. Their great simplicity has been the subject of controversy for many years, some morphologists interpreting it as the result of extensive reduction from more complex ancestors. Others accept it as a sign of great primitiveness.

Two species of *Psilotum* are known, *P. nudum* (= *P. triquetrum*) and *P. flaccidum* (= *P. complanatum*), of which the first is widespread throughout the tropics and subtropics extending as far north as Florida and Spain and as far south as New Zealand. Most commonly, it is to be found growing erect on the ground or in crevices among rocks, but it may also grow as an epiphyte on tree-ferns or among other epiphytes on the branches of trees. *P. flaccidum* is a much rarer plant, occurring in Jamaica, Mexico and a few Pacific Islands, and is epiphytic with pendulous branches.

The organs of attachment in both species are colourless rhizomes which bear numerous rhizoidal hairs and which, in the absence of true roots, function in their place as organs of absorption. In this, they are probably aided by a mycorrhizal association with fungal hyphae, that gain access to the cortex through the rhizoids. Apical growth takes place by divisions of a single tetrahedral cell which is prominent throughout the life of the rhizome, except when dichotomy is occurring. It is said (Bierhorst, 1954b) that this follows upon injury to the apical cell as the rhizome pushes its way through the soil and that two new apical cells become organized in the adjacent regions. In any case there is no evidence of a median division of the original apical cell into two equal halves; to this extent, therefore, the rhizome cannot be said to show true dichotomy.

In *Psilotum nudum*, some branches of the rhizome turn upwards and develop into aerial shoots, commonly about 20 cm tall, but as much as 1 m tall in favourable habitats. Except right at the base, these aerial axes are green and bear minute appendages, usually described as 'leaves', despite the fact that they are without a vascular bundle (cf. *Psilophyton*). The axes branch in a regular dichotomous manner and the distal regions are triangular in cross-section (Fig. 7A). In the upper regions of the more vigorous shoots, the leaves are replaced by fertile appendages (Fig. 7B) whose morphological nature has been the subject of much controversy. Some have regarded them as bifid sporophylls, each bearing a trilocular sporangium, but the interpretation favoured here is that they are very short lateral branches, each bearing two leaves and terminating in three fused sporangia.

Psilotum flaccidum differs from *P. nudum* in two important respects: its aerial branches are flattened and there are minute leaf-traces which, however, die out in the cortex without entering the leaves.

The internal anatomy of the rhizomes varies considerably, according to their size, for those with a diameter of less than 1 mm are composed of almost pure parenchyma, while larger ones possess a well-developed stele. Fig. 7C is a diagrammatic representation of a large rhizome, as seen in transverse section. In the centre is a solid rod of tracheids with scalariform thickenings. As there is no clear distinction between metaxylem and protoxylem it is impossible to decide whether the stele is exarch, mesarch or endarch. Around this is a region which is usually designated as phloem, although it is decidedly unlike the phloem of more advanced plants, for its elongated angular cells are often lignified in the corners. Surrounding this is a region of 'pericycle', composed of elongated parenchymatous cells, and then comes an endodermis with conspicuous Casparian strips in the radial walls. Three zones may often be distinguished in the cortex, the innermost of which is frequently dark brown in colour because of abundant deposits of phlobaphene (a substance formed from tannins by oxidation and condensation). The middle cortex consists of parenchymatous cells with abundant starch grains, while the outer cortex contains, in addition, the hyphae of the mycorrhizal fungus. In some cells the mycelium is actively growing while in others it forms amorphous partially digested masses.

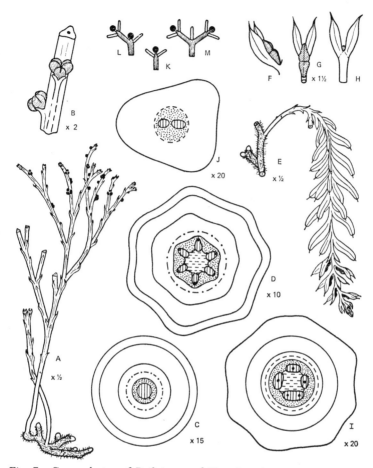

Fig. 7   Sporophytes of Psilotum and Tmesipteris

A–D, *Psilotum nudum*: A, portion of plant showing erect habit;
B, fertile region; C, t.s. rhizome; D, t.s. aerial shoot.
E–J, *Tmesipteris tannensis*: E, portion of plant showing pendulous
habit. F–H, fertile appendages viewed from different directions;
I, t.s. aerial shoot; J, t.s. distal region of shoot. K, theoretical
interpretation of sporangial apparatus of Psilotales. L, M, abnormal
types of sporangial apparatus.

(A, after Bold; B, E–H, Pritzel.)

In the colourless, or brown, transitional region at the base of the aerial axes, the xylem increases in amount, becomes medullated and splits up into a variable number of separate strands. This process of medullation continues higher up the stem, as shown in Fig. 7D, and the central pith region becomes replaced by thick-walled fibres. There is here a transition from the protoxylem, with its helical or annular helical thickenings, to scalariform metaxylem tracheids, the protoxylem being exarch. The xylem is surrounded by a region of thin-walled cells, not clearly separable as phloem and pericycle, and the whole stele is enclosed in a well marked endodermis. The cortex is again divisible into three regions, the innermost containing phlobaphene, the middle region being heavily lignified and the outermost being photosynthetic. The chlorophyllous cells in this outermost region are elongated and irregularly 'sausage shaped', with abundant air spaces between them, which connect with the stomata in the epidermis (Ford, 1904). The leaves are arranged in a roughly spiral manner in which the angle of divergence is represented by the fraction $\frac{1}{3}$, but although internally they are composed of chlorophyllous cells like those of the outer cortex of the stem, they can contribute little to the nutrition of the plant, for they are without stomata as well as having no vascular supply.

In this last respect, the leaves are in marked contrast to the fertile appendages, for these each receive a vascular bundle, which extends to the base of the fused sporangia, or even between them. In their ontogeny, too, they are markedly different from the leaves, for they grow by means of an apical cell, whereas the young leaf grows by means of meristematic activity at its base (Bierhorst, 1956). Shortly after the two leaves have been produced on its abaxial side, the apex of the fertile appendage ceases to grow and three sporangial primordia appear. Each arises as a result of periclinal divisions in a group of superficial cells, the outermost daughter cells giving rise, by further divisions, to the wall of the sporangium, which may be as much as five cells thick at maturity. The inner daughter cells provide the primary archesporial areas, whose further divisions result in a mass of small cells with dense contents. Some of these disintegrate to form a semi-fluid tapetum, in which are scattered groups of spore-mother cells, whose further division by meiosis gives rise to tetrads of cutinized spores.

The genus *Tmesipteris* is much more restricted in its distribution than *Psilotum*, for *T. tannensis* is known only from New Zealand,

Australia, Tasmania and the Polynesian Islands, while another species, *T. Vieillardi*, is probably confined to New Caledonia. (Some workers recognize a further four species, of restricted distribution, although it is possible that they warrant no more than subspecific status.) *T. tannensis* most commonly grows as an epiphyte on the trunks of tree-ferns or, along with other epiphytes, on the trunks and branches of forest trees, in which case its aerial axes are pendulous, but occasionally it grows erect on the ground. By contrast, *T. Vieillardi* is more often terrestrial than epiphytic. It may further be distinguished by its narrower leaves and by certain details of its stelar anatomy.

Like *Psilotum*, *Tmesipteris* is anchored by a dichotomous rhizome with rhizoidal hairs and mycorrhizal fungus hyphae. The aerial axes, however, seldom exceed a length of 25 cm and seldom branch or, if they do so, then there is but a single equal dichotomy. Near the base, the aerial axes bear minute scale-like leaves very similar to the leaves of *Psilotum*, but elsewhere the branches bear much larger leaves, up to 2 cm long, broadly lanceolate and with a prominent mucronate tip (Fig. 7E). Their plane of attachment is almost unique in the plant kingdom, for they are bilaterally symmetrical, instead of being dorsiventral. They are strongly decurrent, with the result that the stem is angular in transverse section and they each receive a single vascular bundle which extends unbranched to the base of the mucronate tip, but does not enter it. In the distal regions of some shoots, the leaves are replaced by fertile appendages which, like those of *Psilotum*, may be regarded as very short lateral branches, each bearing two leaves and terminating in fused sporangia (normally two) (Figs. 7F–H).

The internal anatomy of the rhizome is so similar to that of *Psilotum* that the same diagram (Fig. 7C) will suffice to represent it. In the transition region of *Tmesipteris tannensis* (Fig. 7I), the central rod of tracheids becomes medullated and splits up into a variable number of strands which are mesarch (in contrast to the exarch arrangement in *Psilotum*). (*T. Vieillardi* differs in having a strand of tracheids that continues up into the aerial axis in the centre of the pith region.) Whereas in the rhizome there is a well-marked endodermis, in the aerial axes no such region can be discerned. Instead, between the pericycle and the lignified cortex, all that can be seen is a region of cells packed with brown phlobaphene. The outer cortex contains chloroplasts, but the epidermis is heavily cutinized and is without stomata. These are restricted to the leaves (and their decurrent bases) which, like the stem, are also covered with a very thick cuticle, but in which are abundant stomata. The leaf-trace has its origin as a branch from one of the xylem strands in the stem and consists of a slender strand of protoxylem and metaxylem tracheids surrounded by phloem. As the stem apex is approached, the number of groups of xylem tracheids is gradually reduced (Fig. 7J), all the tracheids being scalariform, even to the last single tracheid.

The vascular strand supplying the fertile appendages branches into three, one to each of the abaxial leaves and one to the sporangial region. The latter branches into three again in the septum between the two sporangia. The early stages of development closely parallel those in *Psilotum* (Bierhorst, 1956), giving rise to thick-walled sporangia containing large numbers of cutinized spores. Both sporangia dehisce simultaneously, by means of a longitudinal split along the top of each.

When discussing the morphological nature of the fertile appendages of the Psilotales, morphologists have made frequent reference to abnormalities (Sahni, 1923) (the study of which is referred to as 'teratology'). In both genera, the same types of variation occur, some of which are represented diagrammatically in Figs. 7L and 7M. The normal arrangement is indicated in Fig. 7K – a lateral axis (shaded) terminating in a sporangial region (black) and bearing two leaves (unshaded). In Fig. 7L one of the leaves is replaced by a complete accessory fertile appendage, while in Fig. 7M both leaves are so replaced and instead of the sporangial region there is a single leaf. There has for a long time been a widely held belief that freaks are 'atavistic', i.e. they are a reversion to an ancestral condition. However, it must be stressed that this belief rests on very insecure foundations. Applying it to the reproductive organs of the Psilotales led to the conclusion that they are reduced from something more complex, at one time assumed to have been a fertile frond. It may well be, however, that the only justifiable conclusion is that, at this level of evolution, leaf and stem are not clearly distinct as morphological categories, and that they are freely interchangeable – interchangeable on the fertile appendages of abnormal plants, just as, on any normal shoot, fertile appendages replace leaves in the phyllotaxy.

The idea that stems and leaves are not fundamentally different and that, in very primitive organisms they grade from one to the other has been developed further by Bierhorst (1968 a and b, 1971, 1973). He visualizes the evolution of compound leaves, like those of ferns, as taking place in two stages. The first involves the appearance of small leaves on a stem and then the second involves the modification of that stem to form the petiole and rachis of a frond in which the small leaves become pinnules. An intermediate stage in the process, therefore, consists of 'non-appendicular fronds' (i.e. leaves which are direct continuations of stems). Bierhorst has observed such intermediates among primitive ferns and he applies a similar interpretation to the aerial shoots of *Psilotum* and *Tmesipteris*. In *Tmesipteris*, the mode of attachment of the leaf to the stem is certainly more like that of a pinna to a rachis, and Bierhorst has found specimens in which the leaves lie in one plane, instead of being arranged spirally round the axis.

On the basis of the similarities that he finds between the Psilotales and such ferns as *Stromatopteris*, *Gonocormus*, *Gleichenia* and *Actinostachys*, Bierhorst suggests that the Psilotales should be suppressed as a taxonomic group and that *Psilotum* and *Tmesipteris* should be transferred to the ferns. However, it should be pointed out that the similarities which Bierhorst demonstrates do not necessarily imply close relationship, and may merely indicate lowly evolutionary status.

Few botanists have had the good fortune to see living specimens of the gametophyte (prothallus) stage of either *Psilotum* or *Tmesipteris*, but all who have testify, not only to their similarity to each other, but also to their remarkable resemblance to portions of sporophytic rhizomes (Holloway, 1917, 1921). So similar are the prothalli and sex organs of *Psilotum* to those of *Tmesipteris* that the same diagrams and descriptions will suffice for both. Like the rhizomes the prothalli are irregularly dichotomizing colourless cylindrical structures, covered with rhizoids (Fig. 8A), and the similarity is further enhanced by the fact that they are also packed with mycorrhizal fungus hyphae. Both archegonia and antheridia are borne together on the same prothallus (i.e. they are monoecious), but because of their small size they cannot be used in the field to distinguish prothalli from bits of rhizomes. Stages in their development are illustrated in Figs. 8B–H (archegonia) and 8I–M (antheridia) (Bierhorst, 1954a).

The archegonium is initiated by a periclinal division in a superficial cell (Figs. 8B and 8C) which cuts off an outer 'cover cell' and an inner 'central cell'. The cover cell then undergoes two anticlinal divisions, followed by a series of periclinal divisions to give a long protruding neck, composed of as many as six tiers of four cells. The central cell, meantime, divides to produce a 'primary ventral cell' and a 'primary neck canal cell' (Fig. 8F). Beyond this stage there are several possible variants, only one of which is illustrated in Fig. 8G, where the primary ventral cell has divided to give an egg cell and a ventral canal cell, while the nucleus of the primary neck canal cell has

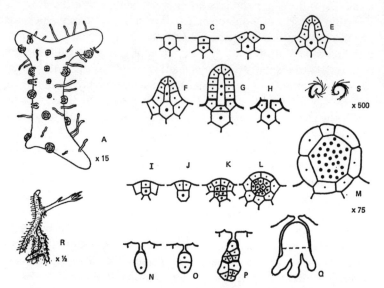

Fig. 8 Gametophytes and embryology of Psilotum and Tmesipteris

A–H, *Psilotum nudum*: A, gametophyte; B–H, stages in development of archegonium. I–S, *Tmesipteris tannensis*: I–M, stages in development of antheridium; N–Q, stages in development of sporophyte; R, young sporophyte attached to prothallus; S, spermatozoids.

(A, Q, S, after Lawson; B–H, Bierhorst; I–P, R, Holloway.)

divided without any cross wall being laid down. In the mature archegonium, however, most of the cells break down so as to provide access to the egg cell from the exterior, through a narrow channel between the few remaining basal cells of the neck, whose walls, in the meantime, have become cutinized (Fig. 8H).

The antheridium, likewise, starts with a periclinal division in an epidermal cell (Fig. 8I). The outer cell is the 'jacket initial' whose further divisions in an anticlinal direction give rise to the single-layered antheridial wall, while the inner 'primary spermatogenous cell' gives rise to the spermatogenous tissue, by means of divisions in many planes (Fig. 8L). At maturity (Fig. 8M) the antheridium is spherical, projects from the surface of the prothallus and contains numerous spirally coiled multiflagellate antherozoids (Fig. 8S). These escape into the surrounding film of moisture and, attracted presumably by some chemical substance, find their way by swimming to the archegonia, where fertilization occurs.

Stages in the development of the young sporophyte from the

fertilized egg are illustrated in Figs. 8N–R. The first division of the zygote is in a plane at right angles to the axis of the archegonium (Fig. 8O) giving rise to an outer 'epibasal cell' and an inner 'hypobasal cell'. The latter divides repeatedly to give a lobed attachment organ called a 'foot' (Fig. 8Q), while the epibasal cell, by repeated divisions, gives rise to the first rhizome, from which other rhizomes and aerial shoots are produced. Fig. 8R shows a young sporophyte with three rhizomatous portions and a young aerial shoot, the whole plant being still attached to the gametophyte. This kind of embryology, where the shoot-forming apical cell is directed outwards through the neck of the archegonium, is described as 'exoscopic'. While relatively unusual in pteridophytes, it is nevertheless universal in mosses and liverworts. Indeed, the young sporophyte of the liverwort *Anthoceros* is very similar indeed to that of *Tmesipteris*, at least up to the stage illustrated in Fig. 8Q, even in such details as the lobed haustorial foot, and some morphologists have gone so far as to suggest some sort of phylogenetic relationship. However, until more is known of the factors which determine the polarity of developing embryos, such suggestions should be received with considerable caution.

For many years there has been speculation among botanists as to the kind of life-cycle that might have been exhibited by the earliest land plants. Some held the belief that there was a regular alternation of sporophytes and gametophytes that resembled each other in their vegetative structure and that even their reproductive organs (sporangia and gametangia, respectively) could be reconciled as having a similar basic organization: on this basis, the generations were regarded as 'homologous'. Others believed that the sporophyte generation evolved after the colonization of the land by gametophytic plants. From being initially very simple, the sporophyte then evolved into something much more complex, by reason of its possessing far greater potentialities than the gametophyte. On this basis, the generations were regarded as 'antithetic'. Until *bona fide* gametophytes are described from the Devonian, or earlier, rocks, there is little hope that this controversy will be resolved satisfactorily. All that can be done is to examine the most primitive living land plants and see whether, at this level of evolution, the sporophyte appears to have fundamentally different capabilities.

The extremely close similarity in external appearance between the gametophytes of the Psilotales and their rhizomes is, therefore, of more than passing interest. Until 1939, however, it was believed that there was one important anatomical distinction between them, in that gametophytes were without vascular tissue. In that year, Holloway (1939) described some abnormally large prothalli of *Psilotum* from the volcanic island of Rangitoto, in Auckland harbour, New Zealand. These were remarkable in having well-developed xylem strands, of annular and scalariform tracheids, surrounded by a region of phloem which, in turn, was enclosed by a clearly recognizable endodermis. There was, therefore, almost no morphological feature distinguishing them from the sporophytic rhizomes, except their archegonia and antheridia. It was subsequently found (Manton, 1942) that the cells of these prothalli contained twice as many chromosomes as those from Ceylon (i.e. they were diploid), while the sporophytes from this locality were tetraploid. To some botanists, this appeared to be sufficient to explain the presence of vascular tissue, and tended to diminish the importance of the similarity of these gametophytes to the rhizomes. But it must be emphasized that diploid prothalli are known elsewhere among pteridophytes and that no morphological aberration need necessarily accompany a simple doubling of the chromsome number. This being so, then, whatever their chromosome content, these abnormal vascularized prothalli still provide strong support for the Homologous Theory of Alternation of Generations. This topic is discussed further in the final chapter.

Concerning chromosome numbers generally in the group, it now appears that all plants of *Psilotum nudum* from Australia and New Zealand have the same chromosome number $n = 100–105$, while plants from Ceylon are like *Psilotum flaccidum* in having about half this number ($n = 52–54$). *Tmesipteris tannensis* has a chromosome number $n = 200+$, while of the six new species (or subspecies) recognized by Barber (1957) five have $n = 204–210$ and one has $n = 102–105$. It is suggested that both *Psilotum* and *Tmesipteris* occur in polyploid series, but that both have the same basic number.

4 Lycopsida

Sporophyte with roots, stems and spirally arranged leaves (microphylls). Protostelic (solid or medullated) sometimes polystelic (rarely polycyclic). Some with secondary thickening. Sporangium thick-walled, homosporous or heterosporous, borne either on a sporophyll or associated with one. Antherozoids biflagellate or multiflagellate.

1 Protolepidodendrales*
    Drepanophycaceae*   *Drepanophycus** (=*Arthrostigma*),
                                   Baragwanathia*
    Asteroxylaceae*   *Asteroxylon**
    Protolepidodendraceae*   *Protolepidodendron**, *Colpodexylon**,
                                  Leclercqia*
2 Lycopodiales
    Lycopodiaceae   *Lycopodites**, *Lycopodium*, *Phylloglossum*
3 Lepidodendrales*
    Lepidodendraceae*   *Lepidodendron**, *Lepidophloios**
    Bothrodendraceae*   *Bothrodendron**
    Sigillariaceae*   *Sigillaria**
    Pleuromeiaceae*   *Pleuromeia**
4 Isoetales
    Isoetaceae   *Nathorstiana**, *Isoetes*, *Stylites*
5 Selaginellales
    Selaginellaceae   *Selaginellites**, *Selaginella*

PROTOLEPIDODENDRALES

At the end of Chapter 2 it was suggested that the Zosterophyllales could conceivably have been the starting point for the evolution of the Lycopsida, and that support for this suggestion was provided by the recently discovered *Kaulangiophyton*. One of the first members of the Lycopsida to appear in the fossil record was *Drepanophycus*, in middle Lower Devonian (Siegenian) rocks of Germany. It was first described as long ago as 1852, and then almost twenty years later some similar remains were described from the Gaspé Peninsula in Canada, under the name *Arthrostigma*. Both names persisted for

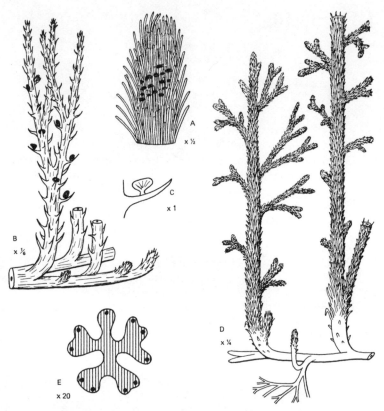

Fig. 9   Drepanophycaceae and Asteroxylaceae

A, *Baragwanathia longifolia*, tip of fertile axis with sporangia shown in
black. B, C, *Drepanophycus spinaeformis*: B, reconstruction;
C, sporophyll. D, E, *Asteroxylon Mackiei*: D, reconstruction;
E, fluted xylem rod, seen in t.s.

(A, based on Lang and Cookson; B, C, after Kräusel and Weyland;
D, after Kidston and Lang.)

some time, but they are now treated as synonyms, of which *Drepanophycus* takes precedence.

Fig. 9B, which is a reconstruction of *Drepanophycus spinaeformis*,
redrawn from Kräusel and Weyland (1935), fails to show the H-
branching that is now known to have occurred in the lower regions
of the plant. Taking this into account, a comparison with Fig. 6D
(*Kaulangiophyton*) shows a striking similarity in the general appearance of the two plants. However, there was a considerable difference

in their size, for the aerial stems of *Drepanophycus* were as much as 5 cm in diameter, and the spine-like outgrowths were as much as 2 cm long. Since each outgrowth had a vascular trace running into it, the term 'leaf' can properly be applied to it. Scattered at random along the aerial axes were 'sporophylls', each consisting of a leaf bearing a single sporangium, either on the adaxial surface (Fig. 9C) or in its axil.

Of the several species of *Drepanophycus* that have been described, *D. spinaeformis* is the best known and also the most widely distributed, having been found in the British Isles, Norway, Siberia and China, as well as in Canada and Germany. Other species have been found in Argentina and in South Africa. *D. spinaeformis* was remarkable not only for its widespread distribution but also for the fact that it is now known to have survived right through the Middle Devonian into the Upper Devonian, having been recently discovered in rocks of Frasnian age in New York State (Banks and Grierson, 1968).

By contrast, *Baragwanathia* is known only from one locality in Australia (Lang and Cookson, 1935). At one time, it was thought not only to be the earliest member of the Lycopsida, but also to be the earliest vascular land plant, for the rocks in which it was found were dated as Silurian. However, they are now believed to be Lower Devonian (either Siegenian or, possibly, Emsian) and, therefore, roughly contemporaneous with the earliest specimens of *Drepanophycus*. *Baragwanathia* (Fig. 9A) had fleshy dichotomizing aerial axes, thickly clothed with leaves, and must have had a most remarkable appearance, for the diameter of the axes ranged upwards from 1 cm to 6·5 cm. In the centre was a slender fluted rod of annular tracheids from which leaf traces passed out through the cortex into the leaves. The leaves were about 1 mm broad and up to 4 cm long and, in fertile shoots, they were associated with reniform sporangia arranged in zones. The preservation of the specimens is not good enough to show whether the sporangia were borne on the leaves or merely among them, but that they were indeed sporangia is established by the extraction of cutinized spores from them. Not much is known of the growth habit of the plant, but there are suggestions that the aerial branches arose from a creeping rhizome.

Asteroxylon Mackiei (Fig. 9D) occurs in the same Scottish deposit as *Rhynia* and *Horneophyton*, but it was much more complex than either of these genera (Kidston and Lang, 1920a). Its aerial axes were about 1 cm across at the base, and they branched monopodially, dichotomous branching being restricted mainly to the lateral branches. In the centre was a deeply fluted rod of tracheids which, in transverse section had a stellate outline (Fig. 9E). Termed an 'actino-stele' by some botanists, it was a solid protostele, consisting entirely of tracheids with either spiral or annular thickenings. The smallest (protoxylem?) elements were near, but not quite at the extremities of the ridges (i.e. mesarch). Surrounding the xylem, was a zone of thin-walled elongated phloem cells. The cortex was composed of three distinct layers, the middle one of which was trabecular (i.e. it consisted of a wide space, crossed by numerous radial plates of tissue), while the innermost and the outermost were of compact parenchyma. Except right at the base, the aerial axes were clothed with enations up to 5 mm long, which were leaf-like in being flattened dorsiventrally and in being provided with stomata. However, they lacked a vascular strand and, for this reason, some morphologists would hesitate to call them leaves. But, in the cortex, there were vascular bundles, which ran from the protoxylems to the bases of the enations and there died out. Because of these, the enations are now usually thought to have represented an early stage in the evolution of leaves. *A. Mackiei* had dichotomous rhizomes whose internal structure was so like that of *Rhynia* that the two were, at first, confused. However, they were remarkable in being completely without rhizoid-al hairs. Instead, small lateral branches of the rhizome grew downwards into the underlying peat, branching dichotomously as they went and it is assumed that they acted as the absorbing organs of the plant.

Associated, but not in organic connection, with the leafy shoots of *Asteroxylon Mackiei* there were some naked forking axes terminating in small pear-shaped sporangia. It was originally thought that these might have been the reproductive organs belonging to *Asteroxylon*, and the plant was accordingly classified along with other members of the Psilopsida that had terminal sporangia. It now seems clear that they belonged to a different plant altogether, to which the name *Nothia* has been given. *Asteroxylon Mackiei* is now known to have borne its sporangia in a lateral position, attached to the axis among the leaves. It is for this reason that it is now classified among the Lycopsida. The sporangia were reniform, and dehisced by means of a terminal slit (Lyon, 1964).

Asteroxylon elberfeldense was described by Kräusel and Weyland (1926) from Middle Devonian rocks near Elberfeld. While the overall height to which *A. Mackiei* grew is not known, the German species is believed to have attained a height of about 1 m. Its internal anatomy differed from that of *A. Mackiei* in that the fluted rod of xylem had a central strand of parenchyma, i.e. it was a medullated protostele.

The original reconstruction of the plant suggested that it grew partially submerged, but this has been questioned by Fairon (1967). Likewise, she has questioned the suggestion that its ultimate branches were naked and bore terminal sporangia. Her interpretation of the plant accords much more closely with that of *A. Mackiei*, as shown in Fig. 9D. However, until its reproductive organs have been positively identified and the manner in which they were borne established, it cannot be said whether *A. elberfeldense* should be classed in the Lycopsida, along with *A. Mackiei*, or whether it should be given a new name and put into a different division of the plant kingdom.

Members of the Protolepidodendraceae are characterized by their having had forked leaves. Thus, the leaves of *Protolepidodendron* were bifid, those of *Colpodexylon* were trifid and those of *Leclercqia* were pentafid. The earliest member of the group to appear was *Protolepidodendron wahnbachense*, from middle Lower Devonian rocks of Germany. *P. scharianum* occurred in Middle Devonian deposits and was more widespread, having been found in China and Australia, as well as in Germany. Other species are recorded from Brazil, Argentina and the USA.

Protolepidodendron scharianum, of which Fig. 10D is a reconstruction, had dichotomous creeping axes, from which arose aerial axes, up to 30 cm tall and less than 1 cm in diameter (Kräusel and Weyland, 1932). All parts of the plant were clothed with leaves that had cushion-like bases. Stems from which the leaves had fallen showed a characteristic pattern of leaf bases, arranged in a spiral manner. Each leaf was provided with a single vascular strand, as was each sporophyll which bore an oval sporangium on its adaxial surface (Fig. 10E). Such sporophylls were scattered among the sterile leaves and were not aggregated into special fertile regions. Details of the stem anatomy of *P. scharianum* are difficult to make out, because of poor preservation, but the stele of *P. gilboense* (Fig. 10F) was a ridged or fluted solid cylinder which, in transverse section, looks like a cogwheel with sixteen teeth, at whose tips the exarch protoxylems were located (Grierson and Banks, 1963).

Two species of *Colpodexylon* have been described (Banks, 1944) one of which (*C. trifurcatum*) is from early Middle Devonian rocks of New York State, and the other (*C. Deatsii*) is from early Upper Devonian rocks of the same region. A reconstruction of the latter species is illustrated in Fig. 10A, which shows the way in which dichotomous aerial axes arose from creeping stems. How tall the plant might have been is not known, but fragments as long as 70 cm have been found. The stems were up to 2·5 cm wide and bore spirally arranged leaves up to 3 cm long, permanently attached to the stem by a cushion shaped base. The vascular structure of the stem was a lobed protostele with exarch or mesarch protoxylems (Fig. 10C), and

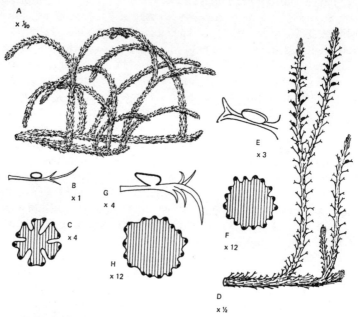

Fig. 10 Protolepidodendraceae

A–C, *Colpodexylon deatsii*: A, reconstruction; B, sporophyll; C, fluted xylem rod, seen in t.s. D–F, *Protolepidodendron*: D, reconstruction of *P. scharianum*; E. sporophyll (bifid, like the sterile leaves); F, fluted xylem rod of *P. gilboense*, seen in t.s. G, H, *Leclercqia complexa*: G, sporophyll (pentafid, like the sterile leaves); H, fluted xylem rod, seen in t.s.

(A–C, after Banks; D, E, Kräusel and Weyland; F, based on Grierson and Banks; G, H, based on Banks, Grierson and Bonamo.)

the metaxylem tracheids had scalariform thickenings. The sporangia were elliptical, about 6 mm long and were attached to the adaxial side of trifid sporophylls (Fig. 10B) scattered among the sterile leaves.

Leclercqia complexa was described by Banks, Bonamo and Grierson (1972) from late Middle Devonian rocks of New York State. It had slender axes, branching in a dichotomous or pseudo-monopodial manner, up to 7 mm in diameter (the longest bit of which so far discovered was 46 cm long). Its leaves were up to 6·5 mm long and, arranged in a low spiral, were attached directly to the stem, there being no leaf cushion. Each had a single vein extending to the tip of the leaf. The xylem strand in the stem was a solid rod with 14–18 protoxylem ridges (Fig. 10H) and the metaxylem was composed of tracheids some of which were scalariform, while others had bordered pits. The sporophylls were like the sterile leaves, in being pentafid, and each had an elliptical sporangium attached to the adaxial surface (Fig. 10G). There were stomata in the sporangium wall, which suggest that, in the living plant, the sporangia would have been green and photosynthetic.

LYCOPODIALES

This group contains two genera of living plants, *Lycopodium* ('Club-mosses') and *Phylloglossum*; and one fossil genus, *Lycopodites*. Of the 200 species of *Lycopodium*, the majority are tropical in distribution, but some occur in arctic and alpine regions. *Phylloglossum*, by contrast, is monotypic and the single species, *P. Drummondii*, is restricted to New Zealand, Tasmania and temperate regions of Australia. Not only do the various species of *Lycopodium* occur in widely different climatic regions; they also occupy widely different habitats, for some are erect bog-plants, others are creeping or scrambling, while yet others are pendulous epiphytes, and this wide range of growth form is paralleled by an extremely wide range of anatomical structure. Indeed, some taxonomists have suggested that the genus should be split into at least four new genera, so different are the various species from one another. Whatever their status, the following sections and subsections of the genus are recognized by most botanists (Engler and Prantl, 1902).

A   Urostachya
1   Selago
2   Phlegmaria

B   Rhopalostachya
1   Inundata
2   Clavata
3   Cernua

Members of the Urostachya never have creeping axes, but have erect or pendulous dichotomous aerial axes, according to whether they are terrestrial or epiphytic. Their roots emerge only at the base of the axis, for although they have their origin in more distal regions, they remain within the cortex (many being visible in any one transverse section of the stem). Perhaps the most important character, phylogenetically, is the lack of specialization of the sporophylls which, as a result, resemble the sterile leaves more or less closely. Another characteristic is that vegetative reproduction may frequently take place by means of bulbils. These are small lateral leafy stem-structures which occur in place of a leaf and which, on becoming detached, may develop into complete new plants. The members of the Rhopalostachya, by contrast, never reproduce by means of bulbils. They are all terrestrial and, although the first formed horizontal axes may be dichotomous, those formed later have the appearance of being monopodial, by reason of their unequal dichotomy, as also do the erect branch-systems. Roots may emerge from the leafy branches, particularly in the creeping parts of the plant.

Of the two sections, the Urostachya (and in particular those belonging to the Selago subsection) are usually regarded as the more primitive. The British species *Lycopodium selago* is illustrated in Fig. 11A. Its sporophylls (Fig. 11B) are very similar indeed to the sterile leaves (Fig. 11C) and occur at intervals up the stem, fertile zones alternating with sterile. *L. squarrosum* shows a slight advance on this, in that the sporophylls are aggregated in the terminal regions of the axes, yet they can hardly be said to constitute a strobilus, for the sporophylls do not differ from the sterile leaves to any marked extent. All the species of the Phlegmaria subsection are epiphytic. *L. phlegmaria* itself is illustrated in Fig. 11K. The pendulous dichotomous branches terminate in branched strobili in which the sporophylls are smaller and more closely packed than the sterile leaves but, nevertheless, afford relatively little protection for the sporangia.

The Inundata subsection of the Rhopalostachya is represented by the British species *Lycopodium inundatum* (Fig. 11D). Here, the strobilus is only slightly different in appearance from the vegetative shoot, for the sporophylls (Fig. 11E) are only slightly modified for protecting the sporangia (cf. sterile leaf, Fig. 11F). Within the Clavata subsection are three more British species, *L. annotinum*, *L. clavatum* and *L. alpinum*, of which the first two are illustrated (Figs. 11G and 11H). In this group, the sporophylls are aggregated into very distinct strobili and are very different from the sterile leaves, for they are provided with an abaxial flange (Fig. 11I) which extends between and around the adjacent sporangia belonging to the sporophylls below (cf. sterile leaf, Fig. 11J). Whereas the strobili of *L. annotinum* terminate normal leafy branches, those of *L. clavatum* are borne on specially modified erect branches, whose leaves are much smaller and more

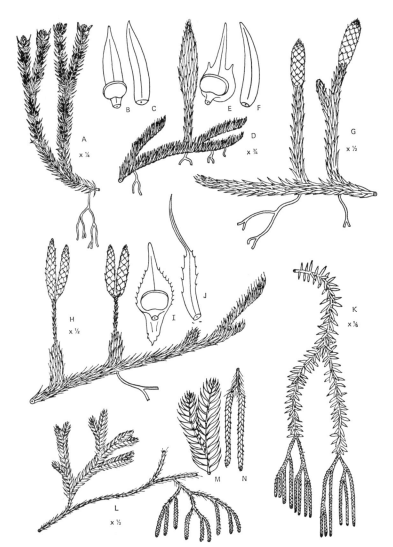

Fig. 11 Lycopodium

A–C, *L. selago*: A, plant; B, sporophyll; C, leaf. D–F, *L. inundatum*:
D, plant; E, sporophyll; F, leaf. G, *L. annotinum*, plant.
H–J, *L. clavatum*: H, plant; I, sporophyll; J, leaf. K, *L. phlegmaria*,
plant. L–N, *L. volubile*: L. plant; M, sterile branch; N, fertile branch.

(B–J, after Hooker; K–N, Pritzel.)

closely appressed. There are, thus, two different kinds of sterile leaf in this species. The Cernua subsection includes a number of species with very different growth habits. *L. cernuum* has a creeping axis, from which arise at intervals erect branch-systems resembling tiny fir trees in being apparently monopodial (for this reason sometimes called 'ground pines'). In this species, all the sterile leaves are alike, but in *L. volubile* (Fig. 11L) there are three or four kinds of sterile leaves. It is a plant with a scrambling habit and its main axes are clothed with long needle-shaped leaves arranged spirally, while the lateral branches are dorsiventral and superficially frond-like. On these branches there are four rows of leaves, two lateral rows of broad falcate leaves (Fig. 11M), an upper row of medium sized needle-like leaves and a row of minute hair-like leaves along the under side. This species, therefore, like several others in this section is highly 'heterophyllous'. The lateral branches in the more distal regions of the plant are fertile and terminate in long narrow strobili, which are frequently branched. As in the Clavata subsection, the closely appressed sporophylls have, on their dorsal (abaxial) side, either a bulge or a flange which provides some protection for the sporangia below.

The apical region of the stem in *Lycopodium* differs markedly from species to species, for it is almost flat in *L. selago*, yet extremely convex in *L. complanatum*. In the past, opinions have differed as to whether growth takes place from an apical cell, but it now appears that this is not the case (Härtel, 1938) and that any semblance of an apical cell is an illusion caused by studying an apex just at the critical moment when one of the surface cells is undergoing an oblique division. All species are now held to grow by means of an 'apical meristem', i.e. a group of cells undergoing periclinal and anticlinal divisions.

The sporelings of all species are alike in their stelar anatomy, for the xylem is in the form of a single rod with radiating flanges. In transverse section these flanges appear as radiating arms, commonly four in number. As the plant grows, the later-formed axes of most species become more complex, the xylem splitting up into separate plates or into irregular strands. However, some species retain a simple stellate arrangement throughout their life, as in *L. serratum* (Fig. 15F) where there are commonly five or six radiating arms of xylem. It is interesting that this species belongs to the Selago subsection which on other grounds is regarded as the most primitive, for some botanists, applying the doctrine of recapitulation, have held that the embryonic structure of a plant indicates what the ancestral condition was like. Such speculations are scarcely necessary, now that the internal anatomy of the earliest fossil members of the Lycopsida has been studied. Without exception, the Protolepido-dendrales can be described as having had a solid fluted rod of xylem, lobed or stellate in transverse section, and the same description applies to members of the Selago section of the genus *Lycopodium*. This section can, therefore, safely be regarded as primitive in its stelar anatomy, as well as in its lack of a well-defined strobilus.

fossil steles

Alternating with the xylem arms in *L. serratum* are regions of phloem, separated from them by parenchyma, and the whole is sur-rounded by parenchymatous 'pericycle', outside which is an endo-dermis. The xylem strand of *L. selago* sometimes shows a slight ad-vance on this arrangement, in that it may be separated into several areas with a variable number of radiating arms. *L. clavatum* has a number of horizontal plates of xylem, alternating with plates of phloem. An even greater number of such plates is found in *L. volubile* (Fig. 15C). To some extent this trend appears to be bound up with an increasing dorsiventrality of the shoot, which reaches its culmina-tion in the heterophyllous *L. volubile*. *L. annotinum* lends some sup-port to this idea, for its horizontal axes are like those of *L. clavatum*, whereas its vertical axes are more like those of *L. selago*. However, exceptions are numerous and it may well be that no valid generaliza-tion of this kind can be made (Jones, 1905).

Quite a different kind of complexity is illustrated by *Lycopodium squarrosum* (Fig. 15E), also placed in the Selago subsection. A trans-verse section of the stem of this species shows not only radiating arms of xylem, but also islands, within the xylem, lined with paren-chyma and containing apparently isolated strands of phloem. Actually, however, the whole structure is an anastomosing one, so that no regions of phloem, or of xylem, are really isolated. This process of elaboration has gone even further in *L. cernuum*, where the appearance is of a sponge of xylem with phloem and parenchyma filling the holes (Fig. 15D).

Throughout the genus, the stele is exarch, the protoxylem elements being clearly recognizable by their 'indirectly attached annular thickenings' (Bierhorst, 1960) (i.e. occasional interconnections occur between adjacent rings), while the metaxylem tracheids are either scalariform or have circular bordered pits. The phloem consists of sieve cells which are elongated and pointed, with sieve areas scattered over the side walls. In young axes it is said that Casparian strips can sometimes be seen in the endodermal cells. In older axes, however, the walls become heavily lignified along with the cells of the inner cortex and their identity becomes obscured. This lignified region extends through most of the cortex in some species, whose stems are consequently hard and wiry, while in other species, e.g. *L. squarrosum*, the stem may be thick and fleshy. Stomata are present in the epidermis of the stem and in the leaves where, in some species, they are on both surfaces ('amphistomatic') and, in others, only on the under side ('hypostomatic'). The leaves of some species are arranged in a whorled or a decussate manner, but in most are spirally arranged. However, in these, the phyllotactic fractions are said to be unlike those of other vascular plants in forming part of the series $\frac{2}{7}$, $\frac{2}{9}$, $\frac{2}{11}$, etc. (Andrews, 1961), whereas the normal phyllotactic fractions, $\frac{1}{2}$, $\frac{1}{3}$, $\frac{2}{5}$, $\frac{3}{8}$, $\frac{5}{13}$, etc., are derived from the Fibonacci series. Each leaf receives a single trace, which has its origin in one of the protoxylems of the stem stele and continues into the leaf as a single unbranched vein composed entirely of spirally thickened tracheids. It is of interest that, in *L. selago*, the bulbils also receive this kind of vascular bundle, for this supports the view that, at this level of evolution, there is no clear morphological distinction between the categories 'leaf' and 'stem'. This is further supported by the fact that leaf primordia may be transformed by suitable surgical treatment into regenerative buds (Williams, 1933).

The so-called 'roots', too, show varying degrees of similarity to stems. All, except the first root of the sporeling, are adventitious and endogenous in origin, arising in the pericycle, and they are peculiar in not bearing endogenous laterals. Instead, they branch dichotomously (very regularly in some species). They are provided with a root cap and their root-hairs are paired (a most peculiar arrangement). The majority are diarch with a crescent-shaped xylem area, but in some species the stele is very similar to that of the stem, as in *Lycopodium clavatum*, where the xylem takes the form of parallel plates.

Variations from species to species in the shape of the sporophylls have already been described. In addition, there is considerable variation in the manner in which the sporangium is borne in relation to the sporophyll. In some, e.g. *Lycopodium selago* and *L. inundatum*, the sporangium is in the angle between the sporophyll and the cone axis, i.e. it is axillary. In others, e.g. *L. cernuum* and *L. clavatum* (Fig. 12C), the sporangium is borne on the adaxial surface of the sporophyll and may be described as 'epiphyllous'. The sporangial initials arise at a very early stage in the ontogeny of the strobilus, normally on the ventral side of the sporophyll, but in some species actually on the axis, whence they are carried by subsequent growth changes into the axil. The first sign of sporangial initials is the occurrence of periclinal divisions in a transverse row of cells (three to twelve in number) (Fig. 12A). The innermost daughter cells provide the archesporial cells by further division and also contribute to the stalk of the sporangium, while the outermost cells (the jacket initials) give rise to the wall of the sporangium (Fig. 12B). This is three cells thick

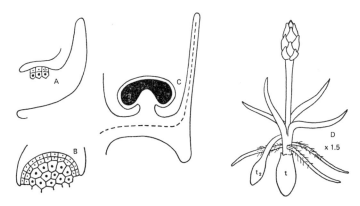

Fig. 12 Lycopodium and Phylloglossum

A–C, *Lycopodium clavatum*, stages in sporophyll development.
D, *Phylloglossum Drummondii*, complete plant. (t, old tuber; t_2, new tuber.)

(C, based on Sykes.)

just before maturity, but then the innermost of the layers breaks down to form a tapetal fluid. Like the sterile leaves, the sporophyll has a single vein, which passes straight out into the lamina, leaving the sporangium without any direct vascular supply. The mature sporangium is kidney-shaped and dehisces along a transverse line of thin-walled cells, so liberating the very numerous and minute spores into the air.

In some species, the spores germinate without delay, while still on the surface of the ground, but in others there may be a delay of many years, by which time they may have become deeply buried. Surface living prothalli are green and photosynthetic, but subterranean ones are, of necessity, colourless and are dependent on a mycorrhizal association for their successful development. Indeed, a mycorrhizal association appears to occur in all species growing under natural conditions, whatever their habit. As a generalization, it may be said that those species inhabiting damp tropical regions germinate rapidly and have green prothalli, whereas those of cooler regions tend to germinate slowly and produce subterranean prothalli. *Lycopodium selago* is interesting in this respect, for it shows variability. Fig. 13A

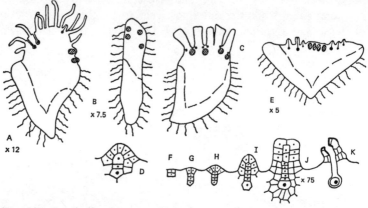

Fig. 13   Prothalli of Lycopodium

A, B, *L. selago*: A, surface-living prothallus; B. subterranean prothallus. C, D, *L. cernum*: C, prothallus; D, archegonium.
E–K, *L. clavatum*: E, prothallus; F–K, stages in development of archegonium.

(A, B, D, I–K, after Bruchmann; E, after Treub.)

illustrates a surface living prothallus with photosynthetic upper regions, in addition to the fungal hyphae in the lower parts (and rhizoids). Fig. 13B, on the other hand, is of a subterranean prothallus, with fungal hyphae in the lower regions but covered all over with rhizoids. Archegonia and antheridia are restricted to the upper parts in both cases. *L. cernuum* provides an example of a surface-living prothallus (Fig. 13C). It is roughly cylindrical and the upper regions bear numerous green photosynthetic lobes, among which are borne the gametangia (Treub, 1884). In *L. clavatum* (Fig. 13E) and *L. annotinum* the prothallus is colourless and subterranean; it is an inverted cone with an irregular fluted margin, growing by means of a marginal meristem which remains active for many years, and the gametangia are developed over the central part of the upper surface. Epiphytic species, e.g. *L. phlegmaria*, also have colourless prothalli, but they are very slender, they branch and they exhibit pronounced apical growth.

In discussions concerning the evolutionary status of the various kinds of prothallus in *Lycopodium*, there has been the suggestion that variability in behaviour, such as that shown by *L. selago*, indicates primitiveness. While this may well be so, it is important to realize that the shape of the prothallus is not a specific character, but one which can be modified by different environmental conditions. Thus, Freeberg and Wetmore (1957) discovered how to break the dormancy of the spores of *L. complanatum*, by treating them with sulphuric acid or by grinding them with sand in dilute detergent; they then showed that prothalli of *L. cernuum*, *L. selago* and *L. complanatum* can be grown on nutrient agar, in the absence of mycorrhizal fungi. Under these conditions, all three species produce almost identical prothalli, with branched chlorophyllous aerial branches. This could be taken to suggest that it is the dormancy of the spores which is significant, rather than the structure and the photosynthetic activity of the prothallus.

Archegonia and antheridia each arise from a single superficial cell in which a periclinal division occurs. The subsequent cell divisions in the antheridial initials are similar to those described for *Tmesipteris* (Fig. 8), but the mature antheridium differs in being sunken into the tissues of the prothallus. The archegonium differs from that of *Tmesipteris* in having several neck canal cells, which vary in number according to whether the prothallus is subterranean or surface living. In the latter species, the neck is very short, e.g. *Lycopodium cernuum* (Fig. 13D), and there may be just a single canal cell, apart from the ventral canal cell. At the other extreme, the number of canal cells may be as high as fourteen in *L. complanatum* (in the Clavata subsection), while *L. selago* is intermediate, with about seven. Various stages in the development of *L. clavatum* are illustrated in Figs. 13F–J. At maturity all the canal cells break down and part of the neck may also wither (Fig. 13K). The antherozoids are pear-shaped and swim by means of two flagella at the anterior end, attracted chemotactically by citric acid diffusing from the archegonium (Bruchmann, 1909 b).

The orientation of the embryo in *Lycopodium* is endoscopic and this is determined at the first division of the zygote, with the laying down of a cross wall in a plane at right angles to the axis of the archegonium (Fig. 14F). The outermost cell, called the 'suspensor', undergoes no further divisions, but the innermost cell gives rise to two tiers of four cells, called the 'hypobasal' and 'epibasal' regions respectively (Fig. 14H). It is from the epibasal (innermost) tier that the young plant is ultimately derived, by further divisions. The hypobasal region remains small in some species, and in others it swells up into a structure commonly called a 'foot' (Bruchmann, 1898). *L. clavatum* is an example of the latter and various stages are illustrated in Figs. 14F–K. In Fig. 14I, the three regions of the embryo are clearly demarcated (the suspensor cell, 's'; the middle hypobasal region, already beginning to swell into a foot, 'f'; the epibasal region

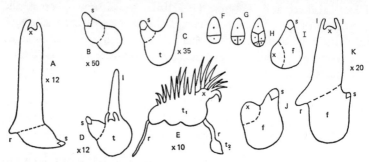

Fig. 14   Embryology of Lycopodium

A, *L. selago*, embryo. B–E, *L. cernuum*, stages in development of embryo. F–K, *L. clavatum*, stages in development of embryo.
(f, foot; l, leaf; r, root; s, suspensor; t, tuber; x, stem apex.)

(A, F–K, after Bruchmann; B–E, Treub.)

with a stem apex, 'x', becoming organized), and the axis of the embryo has bent through a right angle. This bending of the axis proceeds further in Fig. 14J and is completed in Fig. 14K, where, by turning through two right angles, the stem apex is pointing vertically upwards. The first root, 'r', is seen to be a lateral organ, not forming part of the axis of the embryo, as indeed is the case in all pteridophytes: not until the level of the seed plants does the root (radicle) form part of the embryonic spindle. *L. selago* (Fig. 14A) is similar, except that the hypobasal region does not swell up into a large foot.

Lycopodium cernuum (Figs. 14B–E) is an example of a very different kind of embryology (Treub, 1890). As in *L. selago*, the hypobasal region remains relatively small, but the organizing of a stem apex is considerably delayed. The epibasal portion breaks through the prothallial tissue and swells out into a tuberous 'protocorm', 't'. Roughly spherical at the start, it is provided with rhizoidal hairs and mycorrhizal fungus. On its upper surface a cylindrical green leaf ('protophyll'), 'l', appears and then, as the protocorm slowly grows, further protophylls appear in an irregular manner. This stage may persist for a long time and secondary protocorms 't₂' may be formed as shown in Fig. 14E. Finally, however, a stem apex 'x' becomes organized and a normal shoot grows out. This type of development has led, in the past, to much speculation as to its phylogenetic significance, for the protocorm was held by some to represent an atavistic survival of an ancestral condition. However, Wardlaw (1955) has offered an alternative explanation, based on the metabolism of the prothallus and young sporophyte in the various species of *Lycopodium*. He suggests that an abnormally high carbon/nitrogen ratio may delay the organization of a stem apex and may lead, also, to a swelling of the tissues, such being expected where mycorrhizal nutrition is supplemented by photosynthesis. On this basis, the protocorm might well be regarded as a derivative and retrograde development, rather than as a sign of primitiveness.

All the present evidence indicates that *L. selago* is the most primitive living species of *Lycopodium*. Unfortunately, we know relatively little about the immediate ancestors of the genus. While there are fossil remains, known as *Lycopodites*, they contribute little to these discussions. No petrified specimens have been found and some of the mummified remains are now known to be those of conifers. Some had well organized strobili; others did not. *Lycopodites stockii*, from the Lower Carboniferous of Scotland, appears to have been heterophyllous, with its leaves in whorls, and to have had a terminal cone as well as scattered sporophylls among the sterile leaves. Clearly, therefore, this species was very different from the modern *L. selago* and, in some respects, was nearer to some members of the Phlegmaria subsection.

The sporophyte of *Phylloglossum Drummondii*, illustrated in Fig. 12D, is never more than about 4 cm high and appears above ground only during the winter months, when it develops a few cylindrical leaves like the protophylls of *Lycopodium cernuum*. The most robust specimens develop, in addition, a single erect stem terminating in a tiny strobilus. During the hot summer months, when the ground is baked hard, all the aerial parts wither and the plant survives this unfavourable season as a tuber. Each year a new tuber is formed (sometimes two or even three) from the apex of a lateral stem-like structure, which grows out and downwards. This parallel with the behaviour of the protocorm of *L. cernuum* (Fig. 14E) has led to the suggestion that *Phylloglossum* exhibits 'neoteny', in being able to produce sporangia while still in an embryonic stage of development. Whatever the truth of this, it would certainly seem that some of its peculiarities are adaptations which enable it to survive adverse environmental conditions as a geophyte. From the morphogenetic point of view, it is possible to see the tuberization as a response to a high carbon/nitrogen ratio, since the prothallus is both photo-synthetic and mycorrhizal. Perhaps all three 'causes' may apply, for they are not incompatible with each other and merely represent different 'grades of causality'.

Chromosome counts for *Phylloglossum* show a haploid number n = about 255, with many unpaired chromosomes at meiosis, suggest-ing a high degree of hybridization in its ancestry. Such a high num-ber is believed by some to be characteristic of primitive plants, and in this connection it is interesting to find that *Lycopodium selago* has a haploid number n = 130, whereas species in other sections of the genus have lower numbers (*L. clavatum* and *L. annotinum* n = 34). But such a belief is justified only as a generalization. High chromo-some numbers may well point to ancient origins in the majority of cases, but not in all, for polyploidy could have occurred at any stage in the evolution of an organism. Whenever it did occur, further evolution would be retarded because of the masking of subsequent mutations. Thus, if it occurred long ago, the ancient condition would have become 'fixed', as may have happened in *L. selago*; whereas, if it had happened recently, it would be possible for an advanced morphological condition to be associated with a high chromosome number, as in *Phylloglossum* perhaps.

By contrast with the Protolepidodendrales and the Lycopodiales, which are homosporous, the three remaining orders of the Lycopsida (Lepidodendrales, Selaginellales and Isoetales) are heterosporous. Another feature that they share is the possession of a ligule, on the basis of which they are sometimes grouped together as the Ligulatae. The ligule is a minute tongue-like membranous process, attached by a sunken 'glossopodium' to the adaxial surface of the leaves and the sporophylls. A study of living heterosporous lycopods shows that it reaches its maximum development while the associated primordium of the leaf or the sporophyll is still quite small. The mucilaginous nature of the cells and the lack of a cuticle have led to the suggestion that the ligule may keep the growing point of young leaves and young sporangia moist, but the fact is that no-one knows its true function. It may even be a vestigial organ whose function has been lost.

LEPIDODENDRALES

The Lepidodendrales, over 200 species of which are known, first appeared in Lower Carboniferous times and reached their greatest development in the Upper Carboniferous swamp forests, in which members of the Lepidodendraceae, Bothrodendraceae and Sigillariaceae were co-dominant with the Calamitales and formed forests of trees 40 m or more in height. The fourth family, Pleuromeiaceae, is represented by a much smaller plant, *Pleuromeia*, from Triassic rocks, and approached more nearly to the modern Isoetales. The Carboniferous genera had stout trunks, some with a crown of branches, others hardly branching at all, but all possessed the same type of underground organs, known collectively as Stigmarian axes. Some species of *Lepidodendron* (e.g. *L. obovatum*, Fig. 16A) showed very regular dichotomies in its crown of branches, but others approximated to a monopodial arrangement because of successive unequal dichotomies. While the trunks and branches of all species of *Lepidodendron* and *Lepidophloios* were protostelic and exarch, there was nevertheless considerable variation in stelar anatomy, from species to species, and from place to place within one individual. Some species had solid protosteles, others medullated protosteles; some had abundant secondary wood produced by a vascular cambium, some had little and others had none at all; in some, the stele of the trunk had secondary wood, while that of the branches lacked it altogether. Thus, *Lepidodendron pettycurense* and *L. rhodumnense* (both Lower Carboniferous species) had solid protosteles, the former having secondary wood in addition, but the latter being without it. *Lepidodendron selaginoides* (= *L. vasculare*), from the Coal Measures, provides an interesting case of partial medullation, for the central region of the axis consisted of a mixture of parenchyma and tracheids, round which was a solid ring of tracheids. The secondary wood of this species was often excentric in its development, as illustrated in Fig. 15B.

As far as is known, nothing comparable to secondary phloem was produced, either in the trunks and branches (Arnold, 1960), or in the underground axes, *Stigmaria* (Frankenberg and Eggert, 1969). The cambium of the Lepidodendrales divided in such a way that most, if not all, of its products were formed from its inner face. It was, therefore, quite unlike the cambium of gymnosperms and angiosperms.

Lepidophloios Wuenschianus, from the Lower Carboniferous of

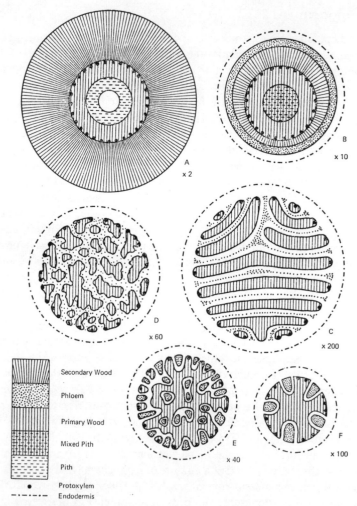

Fig. 15   Steles of various Lycopsids

A, *Lepidophloios Wuenschianus*. B, *Lepidodendron selaginoides*.
C, *Lycopodium volubile*. D, *L. cernuum*. E, *L. squarrosum*.
F, *L. serratum*.
(Leaf-traces have been omitted, for the sake of clarity.)

(C, D, after Pritzel; E, Jones.)

Arran, is known in considerable detail, for examples have been found in which portions of the stele from various levels had fallen into the rotted base of the trunk before petrifaction occurred. This has made it possible to discover something about the growth processes taking place in the young aerial stem. Near the base the primary wood was solid and only 5·5 mm across, half way up the trunk it was medullated, while near the top (Fig. 15A) it was 15 mm across and had a hollow space in the centre of the medulla. It is concluded that, as the stem grew, its apical meristem grew more massive and laid down a much broader procambial cylinder. Meantime, the cambium in the lower regions had laid down more secondary wood than higher up, with the result that the total diameter of the wood (primary and secondary together) was about the same throughout the length of the trunk (about 7 cm). In proportion to the overall diameter of the trunk (40 cm), however, this quantity of wood is surprisingly small, when compared with that of a dicotyledonous tree, where most of the bulk is made up of wood. The difference probably lies in the fact that the wood of modern trees is concerned with two functions, conduction and mechanical support, whereas the wood of Lepidodendrales was concerned only with conduction. Mechanical support was provided mainly by the thick woody periderm which was laid down round the periphery of the trunk.

The metaxylem was composed of large tracheids with scalariform thickenings, while the protoxylem elements were much smaller and frequently had spiral thickenings. The secondary wood consisted of radial rows of scalariform tracheids and small wood-rays, through which leaf-traces passed on their way out from the protoxylem areas. In most specimens the phloem and even some of the cortex had decayed before petrifaction occurred, but what is known of the phloem suggests that it was small in amount and very similar to that of modern lycopods.

The primary cortex was relatively thin-walled and, within it, a number of different regions are recognizable. Of these, the most interesting is the so-called 'secretory tissue', made up of wide thin-walled cells, whose horizontal walls became absorbed in the formation of longitudinal ducts. Each leaf-trace, as it passed through this region, acquired a strand of similar tissue which ran parallel with it before splitting into two 'parichnos strands' on entering the base of the leaf. It is believed that the secretory tissue was in some way connected with the aeration of the underground organs, providing an air path from the stomata of the leaves, through the mesophyll to the parichnos strand and so to the secretory zone, which was

c

continuous with a similar region in the cortex of the Stigmarian axes.

The leaves were borne in a spiral with an angle of divergence corresponding to some very high Fibonacci fraction such as $\frac{55}{144}$, $\frac{89}{288}$, etc. They were linear, up to 20 cm long, triangular in cross-section and with stomata in two longitudinal grooves on the abaxial side. The vascular strand remained unbranched as it ran the length of the leaf. The leaves were shed from the trunk and larger branches by means of an absciss layer, and the shape of the remaining leaf base and scar provides important details for distinguishing the various genera and species. Fig. 16B shows the appearance of the trunk of a *Lepidodendron* where, characteristically, the leaf bases were elongated vertically. In some species, the leaf bases became separated slightly as the trunk increased in diameter, but in others they remained contiguous, even on the largest axes. No doubt this was brought about to some extent by an increase in the size of the leaf base, much as a leaf scar becomes enlarged on the bark of many angiospermous trees, but such increase must have been relatively slight, for otherwise the leaf bases would have become much broader in proportion to their height. Evidently, therefore, the largest leaf bases must have been large from the start, from which it follows that the axes bearing them must also have been large, even when young (Eggert, 1961). Details of a typical *Lepidodendron* leaf base are illustrated in Fig. 16C. Within the area of the leaf scar (2) are to be seen three smaller scars, representing the leaf-trace (3) and the two parichnos strands (4). Above this lies the ligule pit (1) and, in some species, below it are two depressions that were once thought to be associated with the parichnos system, but are now known to be caused by shrinkage of thin-walled cells within the leaf cushion.

Lepidophloios is distinguished by its leaf bases being extended horizontally, instead of vertically. Otherwise, the anatomy of the trunks is indistinguishable from that of *Lepidodendron*. Indeed, it has been suggested that the differences do not warrant a separation into two genera. However, there were differences, in the way the cones were borne. In *Lepidodendron*, they were nearly always terminal, whereas in *Lepidophloios*, they were borne some distance behind the branch tip in a cauliflorous manner.

The cones of both genera are known as *Lepidostrobus* and they consisted of a central axis around which sporophylls were arranged in a compact spiral, their apices overlapping so as to protect the sporangia. Further protection was afforded by a dorsal projection, or 'heel', as illustrated in the idealized longitudinal section, Fig. 16D.

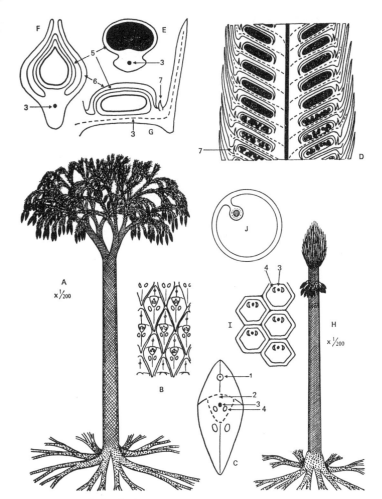

Fig. 16   *Lepidodendrales*

A–C, *Lepidodendron*: A, reconstruction; B, C, leaf bases.
D, E, *Lepidostrobus*: D, l.s. of an idealized cone; E, t.s. sporophyll.
F, G, *Lepidocarpon*: F, t.s.; G, l.s. H, I, *Sigillaria*: H, reconstruction;
I, leaf bases. J, t.s. stigmarian rootlet.
(1, ligule pit; 2, area of leaf base; 3, vascular bundle; 4, parichnos
scars; 5, sporangium wall; 6, flange of sporophyll; 7, ligule.)

(A, C, H, I, after Hirmer; F, based on Scott; G, based on Hoskins
and Cross.)

The cones varied in length from 5 cm to over 40 cm and must have looked like those of modern conifers. Some cones contained only megasporangia, others only microsporangia, while others were hermaphrodite. In the latter, the megasporophylls were at the base and the microsporophylls towards the apex, as illustrated in Fig. 16D. This is the reverse of the arrangement in gymnosperms and angiosperms, where the microsporangial organs lie below the mega-sporangial whenever they happen to be associated in a hermaphro-dite 'flower'. The sporangia of *Lepidostrobus* were elongated and attached throughout their length to the 'stalk' of the sporophyll, which was relatively narrow, compared with the expanded apex of the sporophyll (Fig. 16E). The sporangium wall was only one cell thick at maturity and dehisced along its upper margin. Megaspores and microspores must have been produced in enormous numbers, for they are extremely abundant in all coal-measure deposits. Some mega-spores have been found with cellular contents, representing the female prothallus, retained within the megaspore wall ('endosporic') as in *Selaginella* today, and occasionally archegonia can be recognized.

prothallus ♀ retained ♀ + spore in spore wall [handwritten marginal note]

The number of megaspores produced within each megasporangium varied considerably from species to species, and in some was res-tricted to one. In *Lepidocarpon* (Figs 16F and 16G) the megaspore was retained in the sporangium, which, in turn, was enveloped by two flanges from the stalk of the sporophyll. The whole structure was shed like a seed from the parent plant and has been regarded by some botanists as actually being a stage in the evolution of a seed. It would be much safer, however, to regard *Lepidocarpon* as merely analogous to a seed, for the sporophyll flanges are quite unlike the integuments of true seeds, except perhaps in function. It is not known whether the microspores germinated within the slit-like 'micropyle' while the megasporophyll was still on the tree, or whether it did so after it had fallen to the ground.

Sigillaria (Fig. 16H) is characterized by the arrangement of its leaf bases in vertical rows (Fig. 16I). It branched much less than *Lepidodendron* or *Lepidophloios* and it bore its cones in a cauliflorous manner. Furthermore, the leaves were much longer, up to 1 m, grass-like and, in some species, had two veins, possibly formed by the forking of a single leaf-trace. Species from the Upper Carboniferous were similar to *Lepidodendron* and *Lepidophloios* in their internal anatomy, having a medullated protostele with a continuous zone of primary wood. Some of the Permian species, e.g. *S. Brardi*, however, showed a further reduction of the primary wood, which was in the form of separate circummedullary strands. This is most interesting, for it represents the culmination of a trend which was also taking place, at the same time, among several groups of early gymnosperms, from the solid protostele, through medullated protosteles (first with mixed pith and then with pure pith) to a pith surrounded by separate strands of primary wood.

From a distance, *Bothrodendron* must have looked very similar to *Lepidophloios*, for it had a stout trunk with a crown of branches covered with small lanceolate leaves and its cones (*Bothrodendro-strobus*) were borne in a cauliflorous manner. It differed, however, in the external appearance of the trunk, for it had circular leaf scars that were almost flush with the surface.

The underground organs of all the genera of Lepidodendrales so far described were so similar that they are all placed in the form genus *Stigmaria*, and many are placed in a single artificial species, *S. ficoides*. The base of the trunk bifurcated once and then immediately again, to produce four horizontal axes, each of which continued to branch dichotomously many times in a horizontal plane. These Stigmarian axes were most remarkable structures in many respects. Thus, even at their growing points, perhaps 10 m from the parent trunk, they were frequently as thick as 4 cm. They bore lateral appendages, commonly called 'rootlets', in a spiral arrangement. These were up to 1 cm in diameter and were completely without root hairs. Internally they show a remarkable resemblance to the rootlets of the modern *Isoetes* in having had a tiny stele separated from the outer cortex by a large space, except for a narrow flange of tissue (Fig. 16J). In origin, they were endogenous, although only just so. The axes on which they were borne were peculiar in being completely without metaxylem. In the centre was either pith or a pith-cavity, round which were protoxylem regions directly in contact with a zone of secondary wood. This consisted of scalariform tracheids interspersed with small wood-rays, but there were also very broad rays (through which the rootlet traces passed) which divided the wood into very characteristic wedge-shaped blocks.

The true nature of Stigmarian axes has long been a problem to morphologists, for although doubtless they performed the functions attributed to roots in higher plants (absorption and anchorage), yet they were different in so many respects from true roots and, at the same time, were so different from the aerial axes that they appear to have belonged to a category of plant organization that was quite unique. Even the nature of the 'rootlets' is open to question, for specimens of Stigmarian axes are known which bore leaf-like appendages instead of rootlets. Once more one is forced to the conclusion that the categories root, stem and leaf have no clear distinction at the lower levels of evolution.

Pleuromeia (Fig. 18A) was a much smaller plant than the other members of the Lepidodendrales, for its erect unbranched stems were little more than 1 m high and 10 cm in diameter. The lower parts of the stem were covered with spirally arranged leaf scars, while the upper parts bore narrow pointed ligulate leaves about 10 cm long, attached by a broad base. The plant was heterosporous and dioecious, and the sporangia were borne in a terminal cone made up of numerous spirally arranged circular or reniform sporophylls. Although early accounts described the sporangia as on the abaxial side of the sporophyll, most morphologists believe this to be an error and it is usually accepted that, as in all other lycopods, they were on the adaxial side. Verification of this must await the discovery of better preserved specimens, for no petrified material has yet been found. For this reason, also, little is known of the internal anatomy of the plant.

Below the ground, *Pleuromeia* was strikingly different from the other members of the Lepidodendrales, for, instead of having spreading rhizomorphs of the *Stigmaria* type, it terminated in four (or sometimes more) blunt lobes. From these were produced numerous slender forking rootlets, very similar anatomically to those of *Stigmaria* and also to those of *Isoetes*. Indeed, *Pleuromeia* is commonly regarded as a link connecting the Isoetales with the Carboniferous members of the Lepidodendrales.

ISOETALES

Apart from the fossil genus *Nathorstiana*, the Isoetales contain only the two living genera *Isoetes* and *Stylites*.

The genus *Isoetes* is worldwide in distribution, some seventy species being known, of which three occur in the British flora and are commonly called 'Quillworts'. *I. lacustris* and *I. echinospora* grow submerged in lakes or tarns, while *I. hystrix* favours somewhat drier habitats. Most of the plant is below the level of the soil, with only the distal parts of the sporophylls visible. These are linear structures from 8 to 20 cm long in *I. lacustris*, but up to 70 cm in some species growing in N. America and in Brazil. They constitute the only photosynthetic parts of the plant and, as in many aquatic plants, they contain abundant air spaces (lacunae). The expanded bases of the sporophylls are without chlorophyll and overlap one another to form a bulb-like structure which surmounts a peculiar organ, usually referred to as a 'corm'. The true morphology of the corm has long been the subject of controversy, for it is obscured by a remarkable process of secondary growth, involving an anomalous cambium. This produces small quantities of vascular tissue from its inner surface and large quantities of secondary cortex towards the outside. This secondary cortex dies each year, along with the sporophylls and roots attached to it, and it becomes sloughed off when the new year's growth of secondary cortex is produced. Vertical growth of the corm is extremely slow, with the result that the body of the plant is usually wider than it is high.

Fig. 17A is a diagrammatic representation of a vertical section through an old plant of *Isoetes*. To the right and left are the shrivelled remains of the previous year's growth, the several sporophyll-traces and root-traces being visible within it. All the rest represents the present year's growth surrounding the perennial central regions. Occupying the centre is a solid protostele, the lower part of which is extended into two upwardly curving arms, so that the overall shape resembles an anchor (Fig. 17B). This is made up of mixed parenchyma and peculiar iso-diametric tracheids with helical thickenings. Towards the outside the tracheids are arranged in radial rows but, nevertheless, they are of primary origin (indeed, some workers hold that the whole of the primary wood is protoxylem). Surrounding this primary wood is a narrow zone of phloem (not shown in Fig. 17A), and outside this is the tissue produced centripetally by the anomalous cambium. This commonly consists of a mixture of xylem, phloem and parenchyma and is described by the non-committal term 'prismatic tissue'. The cambium, represented in Fig. 17A as a broken line, cuts through the sporophyll-traces and root-traces of previous years, leaving their truncated stumps still in contact with the primary wood.

What little vertical growth there is takes place by means of apical meristems at the top and bottom of the corm. The lower of these is extended as a line beneath the anchor-shaped primary xylem and is buried deeply in a groove. Roots arise endogenously along the sides of this groove in a very regular sequence and are carried round on to the undersides of the newly formed cortex. The stem apex is also deeply sunken between the 'shoulders' of the corm and is said to contain a group of apical initial cells. Sporophylls arise in spiral sequence (with a phyllotaxy of $\frac{3}{8}$, $\frac{5}{13}$ or $\frac{8}{21}$ in mature plants) and, as new secondary cortex is formed, they are carried up on to the shoulders.

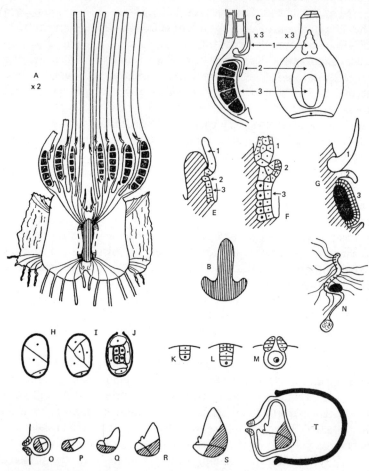

Fig. 17   Isoetes

A, l.s. old plant of *Isoetes* (semi-diagrammatic). B, l.s. stele (at right-angles
to A). C, l.s. leaf base. D, adaxial view of leaf base. E–G, stages in
development of leaf base. H–J, stages in development of male prothallus.
K–M, stages in development of archegonium. O–S, stages in
development of young sporophyte. T, l.s. megaspore, with female
prothallus and young sporophyte.
(1, ligule; 2, velum; 3, sporangium.)

(A–D, based on Eames; E, F, Bower; H–J, Liebig; K–M, Campbell;
N, Dracinschi; O–T, La Motte.)

Stages in the development of the young sporophylls are illustrated in Figs 17E–G. At a very early stage, when the primordium is only a few cells high, one conspicuous cell on its adaxial surface undergoes a periclinal division to produce a ligule primordium (1). This soon gives rise to a membranous ligule a few mm long which, for a time, is much larger than the young sporophyll. Next, a velum initial appears, from which is developed the velum (2) – a flange of tissue which partly hides the sporangium in the mature sporophyll, except for an oval opening called the 'foramen'. The sporangium (3) arises as the result of periclinal divisions in a group of superficial cells near the base of the sporophyll, on its adaxial side. The inner daughter cells are potentially sporogenous, while the outer (peripheral) cells give rise to the sporangium wall, three or four cells thick. *Isoetes* is peculiar among living plants in that some of the potentially sporogenous cells become organized into trabeculae of sterile tissue which cross the sporangium in an irregular manner. They subsequently become surrounded by a tapetal layer which is continuous with the one derived from the innermost layer of the sporangium wall.

The general appearance of the base of a mature sporophyll is indicated in Figs. 17C and 17D, representing a longitudinal section and an adaxial surface view respectively. The sporangia of the Isoetales are larger than those of any other living plant and have a very high spore content indeed. The sporophylls formed earliest in the year and which, therefore, lie outermost on the apex of the corm are megasporangial and contain several hundred megaspores. Those formed later are microsporangial and are estimated to contain up to a million microspores each. Finally, a few sporophylls with abortive sporangia are produced late in the season.

There is no special dehiscence mechanism and the spores are released only when the sporophylls die and decay, as they become sloughed off at the end of the season. The first cell division within the microspore is an unequal one which cuts off a small 'prothallial cell'. The other cell is called the 'antheridial cell' since, by successive divisions (Figs. 17H and 17I), it gives rise to a jacket of four cells surrounding a central cell from which four antherozoids are formed (Fig. 17J). These are spiral and multiflagellate (Fig. 17N) and are released by the cracking of the microspore wall. As already mentioned, this mode of development, where the prothallus is retained within the spore wall, is described as 'endosporic'.

The female prothallus likewise is endosporic. Within the megaspore, free nuclear divisions take place for some time, i.e. nuclei continue to divide without any cross-walls being laid down between them. Then, when about fifty such nuclei have become distributed round the periphery of the cytoplasm, cross-walls are slowly formed, starting in the region immediately beneath the tri-radiate scar, but gradually spreading throughout. Meanwhile, the megaspore wall ruptures at the tri-radiate scar and an archegonium is formed in the cap of cellular tissue which is thereby exposed. Stages in the development of the archegonium are illustrated in Figs. 17K–M. If fertilization does not occur immediately, further archegonia may develop among the rhizoids that cover the apex of the gametophyte.

Stages in the development of the young sporophyte are illustrated in Figs. 17O–T, in which the megaspore is supposed to be lying on its side, as is commonly the case. The first division of the zygote is in a plane at right angles to the axis of the archegonium, or slightly oblique to it. That part of the embryo formed from the outermost half, designated 'the foot', is indicated in the figures by oblique shading. As growth proceeds, the orientation of the embryo changes so that the first leaf and the stem apex are directed upwards, while the first root is directed obliquely downwards. If is of interest that there is no quadrant specifically destined to produce a stem apex, and that it appears relatively late in a position somewhere between the first leaf and the first root. In some species, there are no clearly defined quadrants at all.

Despite the absence of a suspensor, the embryology of *Isoetes* may be described as endoscopic, since it is from the inner half of the dividing zygote that the shoot is ultimately formed.

For some time, the young embryo continues to be enclosed within a sheath of prothallial tissue which grows out round it, but ultimately the various organs break through and the first root penetrates the soil.

A chromosome count of n = 10 has been obtained in one species of *Isoetes*, and of n = 54–56 in several others.

Isoetes is clearly a remarkable genus, not only in its peculiar method of secondary thickening, but also in the fact that all its leaves are, at least potentially, sporophylls. For this reason, some morphologists regard the upper half of the corm as representing a cone axis. The lower half they regard as a highly reduced rhizomorph, homologous with Stigmarian axes, and this is supported, not only by the regular arrangement of the roots on the corm, but also by the extraordinary similarity of the roots to Stigmarian rootlets internally. If this view is correct, then the stem, as such, must have become completely suppressed, along with its leaves.

Stylites was unknown until 1954, when it was first discovered, forming large tussocks round the margins of a lake at an altitude of 4750 m in the Peruvian Andes. Since then, it has been examined in great detail by Rauh and Falk (1959), who claim that there are two species. *Stylites* is no less remarkable than *Isoetes*, for it likewise exhibits a kind of anomalous secondary thickening, though less active, and all its leaves are potential sporophylls. It differs from *Isoetes*, however, in having limited powers of vertical growth and in being able to branch, both dichotomously and adventitiously, so as to form the characteristic tussocks. Two plants are illustrated in longitudinal section, one young and unbranched (Fig. 18C), the other older and branching (Fig. 18D). Perhaps the most remarkable feature is the way in which the roots are borne up one side only and receive their vascular supply from a rod of primary wood which is quite distinct from that supplying the sporophylls; the two run side by side within the axis (Fig. 18E). The nature of the axis is, therefore,

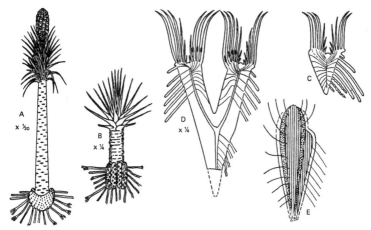

Fig. 18   *Pleuromeia, Nathorstiana and Stylites*

A, *Pleuromeia*, reconstruction. B, *Nathorstiana*, reconstruction. C–E, *Stylites*: C, ls young plant; D, ls older plant; E, ls stele.

(A, after Hirmer; B, Mädgefrau; C–E, Rauh and Falk.)

even harder to interpret than in *Isoetes*. Rauh and Falk draw a comparison with the Cretaceous *Nathorstiana* (Fig. 18B) in which the roots arise from a number of vertical ridges round the base of the stem. This in turn may be compared with *Pleuromeia* and ultimately, therefore, with the Lepidodendrales.

SELAGINELLALES

This group contains two genera, one living (*Selaginella*) and one fossilized (*Selaginellites*). More than 700 species of *Selaginella* are known, of which some occur in temperate regions, but the vast majority are confined to the tropics and subtropics, where they grow in humid and poorly illuminated habitats, such as the floor of rainforests. Some, however, are markedly xerophytic, inhabiting desert regions, and are sometimes called 'resurrection plants' because of their extraordinary powers of recovery after prolonged drought. Relatively few are epiphytes, unlike *Lycopodium*. Some form delicate green mossy cushions, others are vine-like, with stems growing to a height of several metres, while many have creeping axes, from which arise leafy branch systems that bear a striking superficial resemblance to a frond.

Hieronymus (in Engler and Prantl, 1902) divided the genus into the following sections and subsections:

A   Homoeophyllum
   1   Cylindrostachya
   2   Tetragonostachya

B   Heterophyllum
   1   Pleiomacrosporangiatae
   2   Oligomacrosporangiatae

The Homoeophyllum section is a small one, consisting of fewer than fifty species, all of which are isophyllous and have spirally arranged leaves. The only native British species, *Selaginella selaginoides* (=*S. spinosa*, =*S. spinulosa*) (Fig. 19A), is a typical example of this kind of organization and is placed in the subsection Cylindrostachya because the spiral arrangement extends also to the fertile regions. Species belonging to the Tetragonostachya subsection differ in that the sporophylls are arranged in four vertical rows, giving the cone a four-angled appearance. All the members of the Homoeophyllum section are monostelic, but *S. selaginoides* is peculiar in that the stele of the creeping region is endarch (Fig. 19I), whereas that of the later-formed axes is exarch (Fig. 19H), as in all other species. According to Bruchmann (1897) there is a limited amount of secondary thickening in the so-called hypocotyl region of this species; this is the only record of cambial activity in the whole genus.

The Heterophyllum section is characterized by a markedly dorsiventral symmetry and by anisophylly, for the leaves are arranged in four rows along the axis, two rows of small leaves attached to the upper side and two of larger ones attached laterally. The fertile regions, however, are isophyllous and the cones are four-angled,

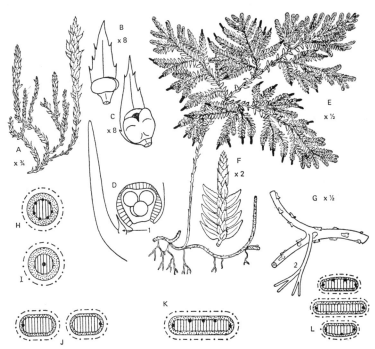

Fig. 19   Selaginella

A–D, *S. selaginoides:* A, whole plant; B, microsporophyll; C, megasporophyll; D, l.s. megasporophyll. E, F, *S. Braunii:* E, portion of plant; F, branch tip with cone. G, *S. Willdenowii*, attachment of rhizophore. H–L, steles: H, *S. selaginoides*, aerial axis; I, *S. selaginoides*, creeping axis; J, *S. Kraussiana*; K, *S. flabellata*; L, *S. Willdenowii*. (1, ligule; 2, rhizophore.)

(A–C, after Hieronymus; H, I, K, Gibson.)

which makes them very clearly distinguishable from the vegetative regions (Fig. 19F). The section is divided, somewhat arbitrarily, on the number of megasporophylls in the cone and is further subdivided on the number of steles in the axis.

Most commonly the axis is monostelic and contains a ribbon-shaped stele, e.g. *Selaginella flabellata* (Fig. 19K), but some species have more complicated stelar arrangements (Gibson, 1894), *S. Kraussiana*, now naturalized in parts of the British Isles, has a creeping habit and has two steles which run side by side (Fig. 19J), except at the nodes where they interconnect. *S. Braunii* (Fig. 19E) is one of the many species which have a creeping stem with erect frond-like branch systems: the creeping axis is bi-stelic, with one stele lying vertically above the other, while the erect axes are monostelic. *S. Willdenowii* is a climbing, or vine-like, species and may have three ribbon-shaped steles (Fig. 19L) or even four. The most complex of all is *S. Lyallii*, where the creeping axis is dicyclic and the aerial axes are polystelic. The central stele of the creeping axis is a simple ribbon, of metaxylem, without any protoxylem, surrounded by phloem, pericycle and endodermis. This is surrounded by a cylindrical stele which is amphiphloic (i.e. has phloem to the inside as well as to the outside of the xylem) and is bounded, both externally and internally, by endodermis. Both steles play a part in the origin of the many steles in the aerial axis, which number twelve or thirteen, four of them being main ones to which the leaf traces are connected, while the rest are accessory steles.

It is important to realize that, however complex the stem of a mature plant may be, the young sporeling *Selaginella* is invariably monostelic, there being a gradual transition along the axis until the adult condition is achieved. This observation has naturally, in the past, led to the supposition that species which are monostelic throughout are more primitive than the more complex species. However, much caution is necessary before accepting this view. In the first place, not all monosteles are directly comparable (e.g. *S. selaginoides* and *S. Braunii*). In the second place, it would seem that some members of the monostelic Homoeophyllum group are highly advanced in other respects. Thus, *S. rupestris* and *S. oregana* (in the Tetragonostachya subsection) are remarkable for the possession of vessels in their xylem. Among the tracheids are lignified elements whose transverse end-walls have dissolved, leaving a single large perforation plate so that, like drain pipes placed end to end, they provide long continuous tubes. While it is true that vessels are known in some other pteridophytes, they are not of this advanced type which occurs, elsewhere, only among the flowering plants. The metaxylem tracheids have scalariform thickenings, while the protoxylem elements are helically thickened and exhibit a feature which is found elsewhere only in *Isoetes* – viz. the helix may be wound in different directions in different parts of the cell (Bierhorst, 1960).

The phloem, composed of parenchyma and sieve cells, is very similar to that of *Lycopodium* and is separated from the xylem by a region of parenchyma one or two cells thick. To the outside of it is a region of pericycle, and then comes a trabecular zone which is characteristic of *Selaginella*. This zone differs markedly in detail from species to species, but is usually a space, crossed in an irregular fashion by tubular cells or by chains of parenchyma cells. Endodermal cells are recognizable also in this region because of their Casparian bands, but it sometimes happens that a single Casparian band may encircle a bunch of several tubular cells. Whatever the exact constitution of the zone, however, it is very delicate, with the result that the stele usually drops out of sections cut by hand. The outer regions of the stem are frequently made up of thick-walled cells and the epidermis is said to be completely without stomata.

In *Selaginella selaginoides*, roots arise in regular sequence from a swollen knot of tissue in the hypocotyl region, but in most creeping species they arise at intervals along the under side of the stem. They are simple monarch structures, which branch dichotomously in planes successively at right angles to each other, as they grow downwards into the soil (Gibson, 1902). Root-caps and root-hairs are present, just as in the roots of other plants. A mycorrhizal association has been demonstrated in *S. selaginoides*. In species with aerial branches, the roots are associated with peculiar organs, usually referred to as 'rhizophores', and some morphologists describe the roots as borne on them, while others describe the rhizophores as changing into roots on reaching the soil. Of these, the second interpretation is probably the more accurate. Rhizophores are particularly well developed in climbing species, such as *S. Willdenowii* (Fig. 19G), where they grow out from 'angle meristems' which occur in pairs, one above and one below, at the junction of two branches. In some species, only one of these is active while the other remains dormant, as a small papilla. The active one grows into a smooth shiny forking structure without leaves. Its branches are without root-caps until they reach the soil, but then root-caps appear and all subsequent branches take on the appearance of typical roots. This is the normal behaviour, but the fate of the angle meristems appears to be under the influence of auxin concentrations, for damage to the adjacent branches may result in their giving rise to leafy shoots, instead of rhizophores. It is clear that the rhizophore fits neither into the category 'stem' nor into the category 'root', but exhibits some of the characters of each. It is not surprising, therefore, that in the days when botanists believed in the reality of these morphological categories, the rhizophores of *Selaginella* were the subject of much argument.

The stem apex shows an interesting range of organization, from species to species, for those with spirally arranged leaves tend to have a group of initial cells, while dorsiventral species usually have a single tetrahedral apical cell (Wand, 1914). Leaf primordia are formed very close to the stem apex and, in some species, appear to arise from a single cell. They give rise to typical microphylls, receiving a single vascular bundle which continues into the lamina as an unbranched vein. The ligule, which is present on every leaf and sporophyll, appears early in their ontogeny and develops from a row of cells arranged transversely across the adaxial surface near the base of the primordium. When fully grown it may be fan-shaped or lanceolate and has a swollen 'glossopodium' sunken into the tissue of the leaf (Gibson, 1896). There is much variation, according to species, in the structure of the lamina of the leaf, for some species possess only spongy mesophyll, while others have a clearly defined palisade layer also (Gibson, 1897). In some, the cells of the upper epidermis and, in others, some of the mesophyll cells contain only a single large chloroplast, a feature which is reminiscent of the liverwort *Anthoceros*. In other species, all the cells of the leaf contain several chloroplasts. There is much variation, also, in the occurrence of stomata, some species being amphistomatic and others hypostomatic.

Early stages in the development of the sporangia in *Selaginella* are very similar to those in *Lycopodium*, and there is a similar range of variation in the location of the primordium. Thus, in some species, it arises on the axis, while in others it arises on the adaxial surface of the leaf, between the ligule and the axis. However, at maturity the sporangium comes to lie in the axil of the sporophyll. The first division is periclinal and gives rise to outer jacket initials and inner archesporial cells. The jacket initials divide further to produce a two-layered sporangium wall and the archesporial cells produce a mass of potentially sporogenous tissue, surrounded by a tapetum. In microsporangia many cells of the sporogenous tissue undergo meiosis to form tetrads of microspores but, in the megasporangia of most species, all the sporogenous tissue disintegrates, except for one spore mother cell, from which four megaspores are formed. Some species, however, retain more than the one functional megaspore mother cell, so that up to twelve or even more megaspores may result. Yet other species are peculiar in that, out of the single tetrad of megaspores, one, two or three may be abortive, so that in the extreme condition the megasporangium may contain only one functional megaspore. *S. rupestris* usually has two megaspores in each sporangium and sometimes only one, while *S. sulcata* regularly has only one. *S. rupestris* is further remarkable in that the megaspores are not shed, but are retained within the dehisced megasporangium and fertilization takes place while it is still *in situ*. Thus, it happens that young sporophytes may be seen growing from the cone of the parent sporophyte. Few seed-plants have achieved this degree of vivipary, yet in *Selaginella* it occurs in a species belonging to the allegedly primitive Homoeophyllum group.

In those species whose cones contain both megasporophylls and microsporophylls, it is usual for the former to be near the base of the cone and the latter near the apex. This further emphasizes the point, already made, that the arrangement in lycopods is the inverse of that observed both in the gymnospermous Bennettitales and in hermaphrodite flowers of angiosperms.

Fig. 19D illustrates the appearance of a megasporophyll in longitudinal section, with the ligule (1) and the differential thickening in the sporangium wall, while Figs. 19B and 19C illustrate a dehiscing microsporangium and megasporangium respectively. Contractions of the thick-walled cells of the megasporangium cause the megaspores to be ejected for a distance of several centimetres, but dispersal of the microspores is mainly by wind currents.

Long before the spores are shed, nuclear divisions have started to take place, so that the prothallus is well advanced when dehiscence occurs. The stages in the formation of the male prothallus are very similar to those figured for *Isoetes* and, at the moment of liberation, the male prothallus commonly consists of thirteen cells (one small prothallial cell, eight jacket cells and four primary spermatogenous cells, of which the latter undergo further divisions to produce 128 or 256 biflagellate antherozoids). Within the megaspore, a large vacuole appears, around which free nuclear divisions occur and then, subsequently, a cap of cellular tissue becomes organized beneath the triradiate scar. In some species, this cap is continuous with the rest of the prothallus, which later becomes cellular too, but in others a diaphragm of thickened cell walls is laid down, as illustrated in Figs. 20F and 20I (d) Rupturing of the megaspore allows the cap to become exposed and it frequently develops prominent lobes of tissue, covered with rhizoids, between which are numerous archegonia. It has been suggested that the rhizoids, as well as anchoring the megaspore, may serve to entangle microspores in close proximity to the archegonia.

The archegonia are similar to those of *Isoetes*, except that the neck is shorter and consists of two tiers of cells only.

The embryology of *Selaginella* is remarkable for the very great differences that occur between species (Bruchmann, 1897, 1909a, 1912, 1913). These are illustrated in Figs. 20A–L, all of which, for ease of comparison, are drawn as if the megaspore were lying on its side. The first cross wall is in a plane at right angles to the axis of the archegonium (Fig. 20A) and the fate of the outermost half in the

Fig. 20   Embryology of Selaginella

A–D, *S. Martensii*, stages in development of embryo. E, *S. selaginoides*, embryo. F–H, *S. Poulteri*. I, J, *S. Kraussiana*. K, *S. Galleottii*. L, *S. denticulata*.

(a, archegonial tube; d, diaphragm; f, foot; r, root; s, suspensor; x, stem apex.)

(All after Bruchmann.)

different species is indicated by oblique shading. In *S. Martensii* (Figs. 20A–D) the outer half gives rise to a suspensor (s), while the inner half gives rise to all the rest of the embryo, with a shoot apex, (x), a root (r) and a swollen foot (f). The axis of the embryo, in this species, becomes bent through one right angle so as to bring the shoot apex into a vertical position. *S. selaginoides* (Fig. 20E) is similar, except for the absence of a foot. *S. Poulteri* (Figs. 20F–H) is a species with a well-developed diaphragm, through which the embryo is pushed by the elongating suspensor. A curvature through three right angles then brings the shoot into a vertical position. *S. Kraussiana* (Figs. 20I and 20J) likewise has a diaphragm, but in this species the venter of the archegonium gradually extends through it (a), so carrying the embryo with it into the centre of the prothallus. The outer half of the dividing zygote provides, not only the vestigial suspensor, but also the foot. The archegonium of *S. Galeottii* behaves in a similar way, but the embryo is different (Fig. 20K) in that the outer half provides the suspensor, the foot and also the root. *S. denticulata* (Fig. 20L) has the various parts of the embryo disposed as in *S. Martensii* (i.e. the root lies between the suspensor and the foot) but they are derived in a completely different way, for they all come from the outermost half of the dividing zygote.

Such extraordinary variations as these are very puzzling and have occupied the thoughts of many morphologists. Some have held that the presence of a well-developed suspensor is a primitive character and that the reduction of this organ in some species is a sign of relative advancement. Its reduction seems to be correlated with the transference of its function to the venter of the archegonium, and this would certainly seem to be an advanced condition. As to the 'foot', all that can be said is that it has little reality as a separate organ, since it can apparently be formed from various regions of the embryo and may even be dispensed with altogether.

Selaginella is peculiar among pteridophytes for its low chromosome numbers, n = 9 being the commonest number, and its chromosomes are minute.

Townrow (1968) has described a fossil species of *Selaginella* (*S. Harrisiana*) from Permian coal measures of Australia, whose aerial parts were closely comparable with those of *S. selaginoides*, differing only in minor details such as the leaf margins. However, its lower rhizomatous parts are less easy to interpret, and Townrow draws an analogy with those of *Tmesipteris*. Unfortunately, nothing is known of its internal anatomy.

Selaginella fraipontii is known in much greater detail. It has been recorded from Lower Carboniferous deposits in Scotland, and from Upper Carboniferous deposits in Belgium and the USA. Its cones were described under the name *Selaginellites crassicinctus* by Hoskins and Abbott (1956), who showed that they contained megaspores which had long been known under the name *Triletes triangularis* as one of the commonest spores in coal-measure deposits. This discovery suggests that *Selaginella* was probably an important component of the flora of those times, contemporaneous with the tree-like Lepidodendrales. Its stems were described under the name *Paurodendron Fraipontii* by Phillips and Leisman (1966), who enlarged upon an earlier description by Fry (1954). Continuity between the various parts of the plant was then demonstrated by Schlanker and Leisman (1969), who at the same time suggested a close relationship to the living *S. selaginoides*.

The basal region of the plant was somewhat swollen, and showed a limited amount of secondary thickening (less than 0·8 mm thick). The aerial portions had a sprawling habit and bore spirally arranged ligulate leaves, and a central exarch protostele, which was stellate in transverse section, with protoxylem arms varying in number from 3–21. The megasporophylls were arranged in the lower regions of the cone and the microsporophylls in the upper.

The elucidation of the detailed morphology and anatomy of this Carboniferous species of *Selaginella* is of the greatest importance, for it confirms the view that of all the living species of *Selaginella*, *S. selaginoides* is the most primitive. It is then of great significance that of all living species of *Selaginella* it is this one which approaches most nearly to the most primitive *Lycopodium* species, *L. selago*. Both are erect and isophyllous, with spirally arranged leaves showing the least difference between fertile and sterile regions and both have simple protostelic vascular systems. The similarities extend even to the young embryo, as a comparison of Figs. 14A and 20E will show. The lack of a well-developed foot in each is interesting, and makes one wonder whether it might have been absent from their ancestors also.

The most important differences between these two plants, therefore, seem to be the heterospory of *Selaginella* and its possession of a ligule. If it be accepted that heterospory is derived from homospory, there remains only the ligule to be explained. This is, indeed, difficult. There is no obvious reason why, in lycopods, this structure should be associated with heterospory. *Selaginella* is usually grouped with *Isoetes* and the Lepidodendrales on the basis of the possession of these two characters, yet on other grounds *Selaginella* stands apart from *Isoetes*. The multiflagellate antherozoids of the latter suggest very fundamental differences. On the basis of the number of flagella, *Lycopodium* and *Selaginella* should be grouped together. Unfortunately, of course, we have no knowledge of the antherozoids of the Lepidodendrales, but one's guess would be that they were multiflagellate, like those of *Isoetes* and *Stylites*. One thing is fairly certain

– that *Selaginella* is not a direct descendant of the Lepidodendrales. Apart from this, one's views on the relationships of the Lycopsida must depend upon a decision as to whether the ligule is more significant phylogenetically than the number of flagella.

5 Sphenopsida

Sporophyte with roots, stems and whorled leaves. Protostelic (solid or medullated). Some with secondary thickening. Sporangia thick-walled, homosporous (or heterosporous), usually borne in a reflexed position on sporangiophores arranged in whorls. Antherozoids multiflagellate.

1 Hyeniales*(?)
   Hyeniaceae*   *Hyenia*,

2 Sphenophyllales*
   Sphenophyllaceae*   *Sphenophyllum*, Sphenophyllostachys*,
                       Bowmanites*, Eviostachya*
   Cheirostrobaceae*   *Cheirostrobus*

3 Calamitales*
   Asterocalamitaceae*   *Asterocalamites*, (= Archaeocalamites*)
   Calamitaceae*   *Protocalamites*(?), Calamites*, Annularia*,
                   Asterophyllites*, Protocalamosatchys*,
                   Calamostachys*, Calamocarpon*, Palaeostachya*

4 Equisetales
   Equisetaceae   *Equisetites*, Equisetum

HYENIALES

Until recently, there were two genera of fossil plants which together comprised the Hyeniales. These were *Hyenia*, of which three species have been described, and *Calamophyton*. However, Leclercq and Schweitzer (1965) have now shown that the internal anatomy of *Calamophyton* was like that of the Cladoxylales. As a result, that genus has now been removed, leaving *Hyenia* as the only representative of the Hyeniales.

Hyenia elegans, from Middle Devonian rocks of Belgium (and also reported from Russia and from one Lower Devonian locality in Germany) is known in greater detail than either of the other two species, *H. sphenophylloides* from Norway and *H. Vogtii* from Spitzbergen. First described by Kräusel and Weyland (1926), it was examined in greater detail by Leclercq (1940), who employed the technique of *dégagement*. This technique involves the careful removal of the rock matrix, grain by grain, so that three-dimensional structures can be followed down into the rock.

Part of the reconstruction published by Leclercq is illustrated in Fig. 21E. There was a stout creeping axis, from which erect axes up to 30 cm tall were produced at intervals, some sterile and others fertile. The sterile ones bore forking appendages which, according to her, were in whorls, alternating at successive nodes. The fertile appendages, of which one is illustrated in Fig. 21F, were likewise borne in whorls, and had reflexed sporangia. It is difficult to decide whether to call them sporangiophores or sporophylls but, until their internal anatomy is known, the term sporangiophore is preferable. Likewise, it cannot be said whether the sterile appendages were leaves or whether they were lateral branches of limited growth performing the photosynthetic functions that one normally associates with leaves.

Schweitzer (1972) has collected further specimens of *Hyenia elegans* in which the erect axes branched up to four or five times. In these specimens he does not find any convincing evidence that the lateral appendages were whorled. Indeed, according to him, they were spirally arranged, except in the mid-regions, where crowding has resulted in apparent verticils. Accordingly, Schweitzer believes that, like *Calamophyton*, *Hyenia* should be transferred to the ferns, and that its nearest relatives were the Cladoxylales. If this were done, then the earliest representatives of the Sphenopsida in the fossil record would be *Sphenophyllum subtenerrimum* and *Eviostachya Hoegii*, both of which are from late Upper Devonian deposits.

SPHENOPHYLLALES

This group first appeared in the Upper Devonian and persisted until the Lower Triassic, remains of stems as well as of leaves being referred to the genus *Sphenophyllum*. More than fifty species are known (Boureau, 1964), all of which are characterized by the whorled arrangement of the leaves, usually in multiples of three at each node (Fig. 21G). The stems were usually very delicate, in spite of secondary thickening, for they seldom exceeded 1 cm in diameter. Presumably, therefore, they were unable to support their own weight and must have been prostrate on the ground, or must have depended on other plants for support. In general appearance, they probably looked rather like a *Galium* ('Bedstraw'). The anatomy of the stem was most peculiar in its resemblance to that of a root, for in the centre was a triangular region of solid primary wood, with the protoxylems at the three corners in an exarch position. In the Lower Carboniferous species, *S. insigne*, the protoxylem tended to break down to form a 'carinal' canal, but in the Upper Carboniferous species this rarely happened. Outside the primary wood, a vascular cambium gave rise to secondary wood, first between the protoxylems, and then later extending all round. However, the wood opposite the protoxylems was composed of smaller cells than on the intermediate radii, resulting in a pattern which is quite characteristic and which is recognizable at a glance in transverse sections (Fig. 21H). The primary wood consisted entirely of tracheids (i.e. without any admixture of parenchyma) and they bore multiseriate bordered pits on their lateral walls. The tracheids of the secondary wood also bore multiseriate pits, but they were restricted to the radial walls. Between the tracheids, there were wood rays. These were continuous in *S. insigne*, but were interrupted in *S. plurifoliatum* where they were represented only by groups of parenchyma cells in the angles between adjacent tracheids. Eggert and Gaunt (1973) have been fortunate enough to find petrified material so well preserved that the tissues immediately external to the secondary xylem can be studied in detail. They have shown that, unlike the cambium of woody pteridophytes such as *Lepidodendron* and *Calamites*, that of *Sphenophyllum* is bi-facial (like that of seed plants), i.e. it gives rise to secondary phloem from its outer face, as well as to secondary xylem from its inner face. Large stems had a considerable thickness of cork on the outside, formed from a deep-seated phellogen.

The leaves of *Sphenophyllum* showed a wide range of structure, some being deeply cleft, while others were entire and deltoid (Figs. 21I–K); yet all received a single vascular bundle, which dichotomized very regularly within the lamina. Some species were markedly heterophyllous, as illustrated in Fig. 21G, and in these the deeply cleft leaves were usually near the base, while the entire ones were higher up on lateral branches, an arrangement that suggests that the former might represent juvenile foliage.

A number of cones, referred to the genera *Sphenophyllostachys* or *Bowmanites*, have been found attached to the parent plant; others, found detached, are placed in these genera on the basis of their general similarity. A number of other genera of cones are also referred to the Sphenophyllales, but on less secure grounds. Some of them represent the most complex cones in the whole plant kingdom. One of the earliest to appear in the fossil record is *Eviostachya*, described

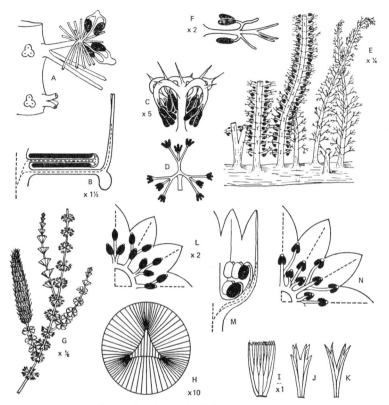

Fig. 21   Hyeniales and Sphenophyllales

A, *Sphenophyllostachys* (= *Bowmanites*) *fertilis*, reconstruction of part
of cone. B, *Cheirostrobus pettycurensis*, sporangiophore and bract.
C, D, *Eviostachya Hoegii*: C, sporangiophore; D, mode of branching
of sporangiophore. E, F, *Hyenia elegans*: E, reconstruction;
F, sporangiophore. G–K, *Sphenophyllum cuneifolium*: G, reconstruction;
H, stele; I–K, leaves. L–N, *Sphenophyllostachys*: L, *S. Dawsoni*, part
of cone in t.s.; M, part of cone in l.s.; N, *S. Roemeri*, part of cone in t.s.

(A, C–F, after Leclercq; B, Scott; G, Smith; I–K, Jongmans;
L, N, Hirmer.)

by Leclercq (1957), from the Upper Devonian of Belgium. Less than 6 cm long and less than 1 cm in diameter, each cone had at its base a whorl of six bracts. Above this were whorls of sporangiophores, six in each whorl. Each sporangiophore was itself highly complicated (Fig. 21C) and branched in a very characteristic way (Fig. 21D), bearing a total of twenty-seven sporangia in a reflexed position. Sporangiophores in successive whorls stood vertically above each other, as is characteristic of the Sphenophyllales, but there were no bracts between them.

Cheirostrobus, from the Lower Carboniferous of Scotland, was a large cone, 3·5 cm across, and had thirty-six sporangiophores in each whorl, subtended by the same number of bracts, each with bifurcated tips (Fig. 21B). The arrangement of the vascular supply to these appendages is interesting in that a common 'trunk-bundle' supplied three sporangiophores and the three bracts subtending them. This has led some morphologists to suggest a more complicated interpretation of the cone structure than is really necessary, based on the supposition that each trunk-bundle represented the vascular supply to one compound organ made up of three fertile leaflets and three sterile leaflets.

Sphenophyllostachys fertilis (= *Sphenophyllum fertile*, = *Bowmanites fertilis*) from the Upper Carboniferous (Fig. 21A) was also a complex cone. Up to 6 cm long and 2·5 cm in diameter, it was made up of whorls of superimposed sporangiophores, six in a whorl, each subtended by a pair of sterile appendages (possibly homologous with one bifid bract). Each sporangiophore terminated in a 'mop' of branches, about sixteen in number, each bearing two reflexed sporangia. Only detached cones have, so far, been found, but they are presumed to have belonged to some member of the Sphenophyllales, because of the triarch or hexarch arrangement of the primary wood in the axis.

Sphenophyllostachys (= *Bowmanites*) *Dawsoni*, on the other hand, is known to have been borne on stems like those of *Sphenophyllum plurifoliatum*. The cone was up to 12 cm long and 1 cm in diameter and bore whorls of 14–20 bracts, fused into a cup near the base, but with free distal portions. In the axils of these bracts, and fused with them to a certain extent (Fig. 21M), were two or three sporangiophores, each with a single reflexed sporangium at its tip. Hoskins and Cross (1943) state that, in most cones of this species, each bract subtended only two sporangiophores and that the arrangement illustrated in Fig. 21L was relatively rare.

Sphenophyllostachys (= *Bowmanites*) *Roemeri* was similar in its organization to *S. Dawsoni*, except that each sporangiophore bore two reflexed sporangia (Fig. 21N).

In recent years, a number of relatively simple cones have been described, which are nevertheless believed to belong to the Sphenophyllales. Thus, *Bowmanites bifurcatus* had only six bifid bracts in each whorl, subtending six sporangiophores, each of which forked once and bore two reflexed sporangia (Andrews and Mamay, 1951). In *Litostrobus iowensis* the number of bracts in a whorl ranged from 6–14. According to Mamay (1954), the bracts were twice as numerous as the sporangia, but more recent work (Baxter, 1967) shows that there was no constant relationship between the number of bracts and the number of sporangia. Both workers agree, however, that the sporangia in this interesting cone were not reflexed. Despite this departure from the typical sphenopsid arrangement, there seems to be little doubt of the affinities of *Litostrobus* because another species, *L. paulus* (=*Mesidiophyton paulus*), has been found attached to stems which are indistinguishable from *Sphenophyllum*.

While the vast majority of the Sphenophyllales were homosporous, at least one, *Bowmanites delectus*, was heterosporous (Arnold, 1947), with megaspores about ten times the size of the microspores.

CALAMITALES

This group reached the peak of its development in the Upper Carboniferous, when a large number of arborescent species was co-dominant with the Lepidodendrales in coal-measure swamp forests; yet by the end of the Permian the group had become extinct. The first representatives to appear, in the Upper Devonian, were the Asterocalamitaceae, a group which differed from the later Calamitaceae in a number of interesting details. *Asterocalamites* (=*Archaeocalamites*) (Fig. 22A) had woody stems up to 16 cm in diameter, strongly grooved on the outside, with the grooves continuing through successive nodes (i.e. not alternating). The leaves, up to 10 cm long, were in whorls at the nodes and forked many times dichotomously, in a manner strongly reminiscent of *Calamophyton* leaves. At intervals along the more slender branches, there were fertile regions, in which there were superimposed whorls of peltate sporangiophores, each bearing four reflexed sporangia (Fig. 22B). Sometimes the fertile regions were interrupted by a whorl of leaves, but these were apparently normal leaves and could not be regarded as bracts. The absence of any regular association between bracts and sporangio-

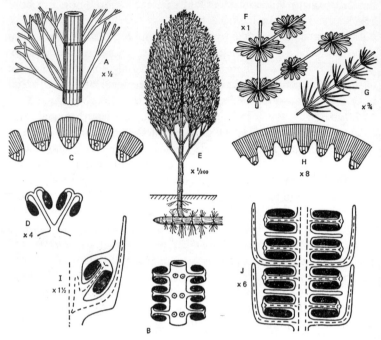

Fig. 22   Calamitales

A, B, *Asterocalamites* (= *Archaeocalamites*): A, stem and leaves; B, fertile region. C, *Protocalamites*, part of internodal vascular system. D, *Protocalamostachys*, sporangiophore.   E–H, *Calamites*: E, reconstruction of *Eucalamites* type; F, *Annularia*-type leaves; G, *Asterophyllites*-type leaves; H, *Arthropitys*, part of internodal vascular system. I, *Palaeostachya*, sporangiophore and bract in l.s. J, *Calamostachys*, part of cone in l.s.

(A, after Stur; B, Renault; D, based on Walton; E, after Hirmer; F, G, Abbott; I, based on Baxter; J, Scott.)

phores makes an interesting comparison with the cones of the later Calamitaceae.

Protocalamites, from Lower Carboniferous rocks of Pettycur, Scotland had ridged stems in which it was stated by Scott (1923) that the ridges alternated at successive internodes. For this reason, it was placed in the Calamitaceae, as the earliest representative of the group. However, in transverse section, its stems show an important anatomical difference for, as illustrated in Fig. 22C, there was a marked development of centripetal wood, as well as centrifugal (i.e. the primary wood was mesarch). As in *Calamites* and in *Equisetum*, the protoxylem tended to break down to form a carinal canal. Secondary wood was laid down to the outside of the meta-xylem, but the primary wood-rays were so wide that it gives the appearance of having been formed in separate strands, although in fact it was formed from a continuous vascular cambium. Chaphekar (1963), having examined further material of *Protocalamites*, concludes that the ridges in successive internodes did not alternate, except very occasionally, and that the similarities between *Protocalamites* and *Archaeocalamites* are so close as to suggest that they are synonymous, and that the former generic name should therefore be suppressed.

Protocalamostachys is the name given to a peculiar cone described by Walton (1949) from Lower Carboniferous rocks in the Island of Arran. Two small pieces of the cone had dropped into the hollow stump of a *Lepidophloios* before it became petrified. Unlike the cones of other members of the Calamitales, its sporangiophores branched twice (Fig. 22D), instead of being peltate. In this respect, it showed some affinities with the Sphenophyllales and also with the Hyeniales. However, Walton compares it most closely with *Pothocites*, a cone associated with leaves of the *Asterocalamites* type. Furthermore within the axis of the cone there was centripetal primary wood as in the stem of *Protocalamites*.

The height to which *Calamites* grew is difficult to determine, because of the fragmentary nature of the remains, but it is almost certain that some specimens must have attained a height of 30 m with hollow trunks whose internal diameter was up to 30 cm. Strictly speaking, the generic name *Calamites* should be applied only to pith casts of stems and branches, while petrified wood should be described under the form genus *Arthropitys*, but common usage has extended the application of the name *Calamites* to include all methods of preservation. The pith casts exhibit ridges and grooves, corresponding in number to the protoxylem strands, running up the inside of the secondary wood and alternating at successive nodes. In this respect they differ from *Mesocalamites*, in which there was some variability from node to node, the ridges sometimes alternating and sometimes continuing straight across the nodes. A number of subgenera are recognized which differed in their mode of branching and, hence, in their general form. The subgenus *Eucalamites* branched at every node. Fig. 22E is of *Eucalamites carinatus*, in which there were only two branches at each node, but other species branched more profusely. By contrast, the subgenus *Stylocalamites* branched only near the top of the erect organpipe-like trunk.

Transverse sections of *Calamites* (*Arthropitys*) show very little primary wood indeed, for secondary thickening provided most of the wood (Fig. 22H). The protoxylem was represented by carinal canals and the small amount of metaxylem present was entirely centrifugal. The wood-rays varied, according to species, dividing the secondary wood into segments in some species, or losing their identity in a continuous cylinder of wood in others. In all cases the wood contained small wood-rays in addition, but otherwise was composed entirely of tracheids with scalariform pitting or with circular bordered pits on the radial walls. Few specimens have been found that are well enough preserved to show details of the tissues immediately external to the cambium. However, Agashe (1964) has examined such specimens, both of stem and of root. In neither is there a sign of anything resembling the secondary phloem that is found in higher plants with secondary thickening. The cells occupying this region are, for the most part, only slightly elongated and they are without sieve areas.

The leaves of *Calamites* were unbranched, with a single mid-vein, and occurred in whorls of four to sixty. In most species, they were free to the base, but in a few they showed some degree of fusion into a sheath. They are placed in one or other of two form genera, according to their overall shape, *Annularia* being spathulate or deltoid (Fig. 22F), while *Asterophyllites* were linear (Fig. 22G). The latter were peculiar in being heavily cutinized, with the stomata restricted to the adaxial surface, suggesting that the branches bearing them may have been pendulous. It is probable, therefore, that the cones were pendulous, too.

The cones of *Calamites* were borne in a variety of ways, in some species singly at the nodes, in others in terminal groups or infructescences or on specialized branches. Many species are known, but most of them are placed in one or other of the two genera *Calamostachys* and *Palaeostachya*. Boureau (1964) lists 27 species of the former and 8 of the latter. As originally defined, these two genera were clearly distinct (as shown in Figs. 22I and 22J) but, in the light of many newly discovered species, Andrews (1961) has questioned whether the distinction is now justified. In both genera there were whorls of peltate, or cruciate, sporangiophores bearing four reflexed sporangia (Figs. 22I and 22J), alternating with whorls of bracts fused into a disc near their point of attachment. Whereas the sporangiophores were in vertical rows, the bracts in successive whorls alternated with one another.

Early accounts assume that there was a constant relationship between the number of bracts in a whorl and the number of sporangio-phores. Thus, *Calamostachys Binneyana*, a cone about 3·5 cm long and 7·5 mm wide, was described as having six sporangiophores in each whorl and twelve bracts. However, Taylor (1967) concludes that there was no such constant relationship, there being from 18–22 bracts per whorl and approximately 12 sporangiophores. Likewise, *C. americana*, a much larger cone, more than 12 cm long and 4 cm in diameter, had between 40 and 45 bracts per whorl and approximately 30 sporangiophores. *C. magnae-crucis* was more complicated, in having alternating vascular bundles in successive internodes within the cone and in having sporangiophores and bracts so numbered that, if 'n' were the number of vascular bundles, then the number of sporangiophores was 2n in each whorl and the number of bracts 3n; the number 'n' could be either seven or eight.

Most species of *Calamostachys* were homosporous. One has recently been shown by Baxter and Leisman (1967) to have had within its sporangia spores which had hitherto been known only as dispersed spores under the name *Elaterites triferens*. These were like the spores of living *Equisetum* in having had elaters attached to them. Some species were heterosporous, e.g. *C. casheana*, whose mega-spores were three or four times the size of the microspores, and *C. americana* where they were about twice the size. *Calamocarpon* is the name given to cones whose organization is like that of *Calamostachys*, but which had achieved a degree of heterospory equivalent to that of *Lepidocarpon*, in that the megasporangium contained only a single megaspore. Furthermore, the cones were monosporangiate, i.e. they were either entirely microsporangial or entirely megasporangial. *Calamocarpon insignis* was originally described by Baxter (1963) as having 12 bracts in a whorl and 12 sporangiophores, but Leisman and Bucher (1971b) have found some cones with 6–10, some with 12 or 14 and some with 16.

Whereas the sporangiophores of *Calamostachys* stood out at right angles to the cone axis, those of *Palaeostachya* stood out at an angle of about 45°, and in some species they appear to have been in the axil of the bract whorl below. Just as in *Calamostachys*, so also in *Palaeostachya*, views have changed as to the relationship between the number of sporangiophores in each whorl and the number of bracts. Thus, *P. vera* was originally thought by Williamson and Scott (1894) to have had 8–10 sporangiophores in each whorl and twice as many bracts, but Hickling (1907) described cones with 16, 18 and 20 sporangiophores and an equal number of bracts. More recently still, Leisman and Bucher (1971a) have found specimens with 18 and 20 sporangiophores but 27, 29 and 30 bracts. They conclude that most calamitalean cones have an approximate 1:1, 3:2 or 2:1 bract–sporangiophore ratio, but emphasize that the number of appendages, as well as their ratio, can vary within a single species. This confirms the conclusion by Browne (1927) that bracts and sporangiophores were inserted on calamite cones as independent systems and can vary in number independently of one another. Perhaps the most interesting feature shown by *Palaeostachya vera* is the course taken by the vascular bundle supplying the sporangiophore. As illustrated in Fig. 211, it travelled up in the cortex of the cone axis to a point about midway between the bracts, and then turned downwards, before entering the stalk of the sporangiophore. To those morphologists who regard vascular systems as highly conservative, this implies that *Palaeostachya* must have evolved from some ancestral form in which the sporangiophore stood midway between the bract whorls, as in *Calamostachys*, and that during the 'phyletic slide' the vascular supply had lagged behind.

The above brief review of the Calamitales brings out some interesting evolutionary trends, which are paralleled very closely in the Lepidodendrales. Thus, the production of increasing amounts of secondary wood was accompanied, in both groups, by a reduction of the primary wood, of which the centripetal metaxylem was the first to go, being replaced either by pith or by a central hollow. At the same time, there was a trend in the fertile regions from a 'Selago condition' to a compact cone, in which the sporangia were protected by overlapping sporophylls in one group and by bracts in the other. Then, having reached their zenith together, both groups became extinct at about the same time.

EQUISETALES

The only representatives of the Sphenopsida that are alive today belong to the single genus *Equisetum* and, of this, only some fourteen species are known. Nine of them occur in the British Isles, where they are known as 'horse-tails'. The genus is distributed throughout the world with the exception of Australia and New Zealand, from which countries it is completely absent. All the species are herbaceous perennials, but there is an interesting range of growth habits, for some are evergreen, while others die back to the ground each year. Early statements that a limited amount of secondary thickening occurs are now discredited, for there is no evidence that a cambium is present in any species. Most species are, therefore, very limited in size; the largest species, *E. giganteum*, has stems up to 13 m long, but since they are only 2 cm thick the plant depends on the surrounding vegetation for its support. The largest British species, *E. telmateia* sometimes attains a height of 2 m and is free-standing in sheltered localities, but most species are much smaller than this and are between 10 and 60 cm tall.

In all species there is a horizontal rhizome from which arise aerial stems that branch profusely in some species (e.g. *Equisetum telmateia,*

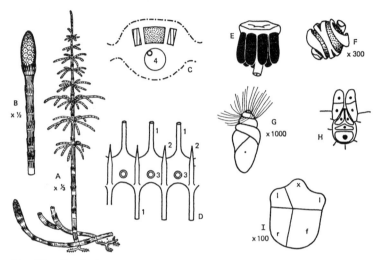

Fig. 23   Equisetum

A, B, *E. pratense*: A, plant with young sterile shoot; B, young fertile shoot with cone. C, *E. sylvaticum*, internodal vascular bundle.
D–G, *E. pratense*: D, nodal vascular arrangement; E, sporangiophore; F, spore, with elaters; G, antherozoid. H, *E. hyemale*, archegonium.
I, *E. arvense*, embryo.
(1, internodal bundles; 2, leaf-traces; 3, branch traces; 4, carinal canal, with remains of protoxylem; f, foot; l, leaf primordia; r, root primordium; x, stem apex.)

(D, H, based on Jeffrey; F. after Foster and Gifford; G, Sharp; I, Sadebeck.)

E. arvense) (Fig. 23A) or remain quite unbranched in others (e.g. *E. hiemale*). The leaves, in all species, are very small and are fused into a sheath, except for their extreme tips which form teeth round the margin of the sheath. They are usually without chlorophyll, photosynthesis being carried out entirely by the green stems. In the past,

D

there have been discussions as to whether the small leaves of *Equisetum* represent a primitive or a derived condition, but, in the light of the fossil record, it is now clear that they have been reduced from larger dichotomous structures (i.e. that they are derived). The stems are ridged, each ridge corresponding to a leaf in the node above, and the ridges in successive internodes alternate with one another (as, of course, do the leaves in successive leaf-sheaths). There are, however, some departures from this regular alternation, as the number of leaves in a whorl diminishes from the base to the apex of the stem (Bierhorst, 1959).

The sporangia are borne in a cone, which in some species (*Equisetum arvense*) terminates a special aerial axis that lacks chlorophyll, is unbranched (Fig. 23B) and appears before the photosynthetic axes. In other species the fertile shoot is green and may subsequently give rise to vegetative branches lower down (e.g. *E. fluviatile* and *E. palustre*), after the cone has withered. In yet other species, most of the lateral branches may terminate in a cone (e.g. the Mexican species *E. myriochaetum*). This last arrangement is commonly regarded as the primitive condition, on the basis that it involves the least specialization, but it must be realized that real evidence for this view is lacking.

The internal anatomy of the stem of *Equisetum* presents an interesting association of xeromorphic and hydromorphic characters, together with a vascular system which is without parallel in the plant kingdom today, and whose correct morphological interpretation has long been the subject of controversy. The ridges in the stem are composed of sclerenchymatous cells, and the epidermis is so heavily silicified as to blunt the edge of the razor when cutting sections. Stomata are restricted to the 'valleys' between the ridges and are deeply sunken into pits whose openings may be partly covered by a flange of cuticle. The walls separating the guard cells from their accessory cells bear peculiar comb-like thickenings which are known elsewhere only in the leaves of *Calamites*. Beneath each of the valleys is a 'vallecular canal' and the central region of the internodes of aerial stems consists of a large space (but, in subterranean stems, the centre may be occupied by pith). At the nodes, there is a transverse diaphragm. Such an arrangement of air channels, together with a very reduced vascular tissue, are features normally found in water plants and contrast strikingly with the heavy cuticle, sunken stomata and reduced leaves.

The internodal vascular bundles lie beneath the ridges of the stem and are quite characteristic (Fig. 23C). As in *Calamites*, the protoxylem is endarch and is replaced by a carinal canal (4), in which may be seen lignified rings which are all that remain after the dissolution of annular tracheids. To the outside of each carinal canal, and on the same radius, lies an area of phloem, flanked on either side by a lateral xylem area. This lateral xylem may contain further protoxylem tracheids with annular thickenings, but otherwise consists of meta-xylem elements which may be tracheids with helical thickening, or with pits, or may even be true vessels. Two types of vessel element occur, one with simple perforation plates and the other with reticulate, but it must be emphasized that they are restricted to the internodes and that they seldom occur more than three in a row. They do not, therefore, form conducting channels of great length as do the vessels of flowering plants (Bierhorst, 1958). In some species (e.g. *Equisetum fluviatile*) each internodal bundle is surrounded by its own separate endodermis, in others (e.g. *E. palustre*) there is a single endodermis running round the stem outside all the bundles, while in yet other species (e.g. *E. sylvaticum*, Fig. 23C) there are two endodermes, one outside and the other inside all the bundles.

At the nodes (Fig. 23D) the vascular bundles (1) are connected by a continuous cylinder of xylem, from which the leaf traces (2) and branch traces (3) have their origin. Vallecular canals occur in this region, but carinal canals are absent, and there has been some disagreement as to whether protoxylem is absent too. However, the most recent investigations confirm its presence as a constant feature (Bierhorst, 1958). This disposes of the view, held by some, that the internodal bundles represent leaf traces extending down through the internode to the node below. An alternative view used to be held – that the vascular network represents a kind of dictyostele, in which the spaces between the internodal bundles represent leaf-gaps. However, this is unlikely, in view of the arrangement known to have existed in the earliest relatives of the genus, such as *Asterocalamites*, where there was no alternation at the nodes. Furthermore, this view overlooked the peculiar way in which the internodes of *Equisetum* are formed from an intercalary meristem. If analogies are to be sought with other pteridophytes, then it is not the internode, but the node, which should be compared. The vascular structure of the node can best be looked upon as a medullated protostele. The internodal spaces then appear as perforations, albeit of a peculiar (intercalary) origin.

Growth at the stem apex takes place as a result of the activity of a single tetrahedral apical cell, daughter cells being cut off in turn from each of its three cutting faces. Despite the spiral sequence of such daughter cells, subsequent growth results in a whorled arrangement and three daughter cells together give rise to all the tissues which make up a node and an internode. It is interesting that, in the first-formed stem of the young sporeling, there are three leaves in each whorl, but, nevertheless, it is stated that their initiation is in no way determined by the position of the cutting faces of the apical cell. Each leaf primordium grows from a single tetrahedral apical cell. In the angle between the leaf sheath and the axis, but on radii between the leaves, lateral bud primordia arise, also with a single apical cell. The lateral bud primordia subsequently become buried by a fusion of the base of the leaf sheath with the axis, with the result that, when it grows, it has to burst through the leaf sheath, so giving the appearance of an endogenous origin. However, not all branch primordia do grow, for in species such as *Equisetum hyemale*, although present, they are inhibited from growing beyond the primordial stage, unless the main stem apex should be destroyed or damaged. Each branch primordium, besides bearing leaf primordia, also bears a root primordium which in aerial axes is also inhibited from growing further. In underground axes, however, they are not inhibited in this way. It is interesting to note that the roots which are apparently borne on a horizontal rhizome are, in fact, borne by the axillary buds hidden within its leaf sheaths, and not directly upon it.

The root grows from an apical cell with four cutting faces, the outermost of which gives the root cap. It may be triarch, tetrarch or diarch in its vascular structure and there is usually just a single central metaxylem element. The stele is surrounded by a pericycle, whose cells correspond exactly in number and radial position to those of the endodermis, since they are formed by a periclinal division in a ring of common mother cells. This has led to the statement that the root has a double endodermis, but this is incorrect, since the cells of the inner ring are without Casparian strips, and must be regarded as pericycle.

The cone (Fig. 23B) invariably terminates an axis, whether it be the main axis or a lateral one, and bears whorls of sporangiophores, without any bracts or other leaf-like appendages interposed, although there is a flange of tissue at the base of the cone called the 'collar'. Each sporangiophore is a stalked peltate structure, bearing five to ten sporangia which, although having their origin on the outer surface, become carried round during growth into a reflexed position on the underside of the peltate head (Fig. 23E). Within the cone axis, the vascular system forms a very irregular anastomosing system, without discernible nodes and internodes, from which the sporangiophore traces depart without any regular association with the gaps.

The sporangiophore trace branches within the peltate head and each branch terminates near a sporangium.

The sporangium has its origin in a single epidermal cell, which divides periclinally into an inner and an outer cell. The inner cell gives rise to sporogenous tissue. The outer cell gives rise to further blocks of sporogenous tissue and also to the wall of the sporangium. Adjacent cells may also add to the sporogenous tissue. The sporangium may therefore be described as eusporangiate in the widest sense and at maturity the sporangium is several cells thick. The innermost wall cells break down to form a tapetum, as also do some of the spore mother cells, and the ripe sporangium wall is two cells thick, of which the outer layer shows characteristic spiral thickenings.

Each spore, as it matures, has deposited round it four spathulate bands, which are free from the spore wall except at a common point of attachment (Fig. 23F). These are hygroscopic, coiling and uncoiling with changes in humidity, and are referred to as 'elaters', although what function they perform during dehiscence of the sporangium is not clear. McClean and Ivimey-Cook (1951) claimed that a distribution curve of the size of the spores in *Equisetum arvense* is a bi-modal one, and that a slight degree of heterospory exists, the large spores being some 25 per cent larger than the small ones. Furthermore, they claimed that the smaller spores give rise to small male prothalli, whereas the larger spores produce hermaphrodite ones. However, Duckett (1970a, 1970b, 1972, 1973) finds no support whatsoever for these claims.

Duckett has grown *in vitro* more than 10000 prothalli, of some ten species of *Equisetum*, in single spore cultures on nutrient agar, free from contamination by fungi or bacteria. The smaller spores mentioned by McClean and Ivimey-Cook are, in fact, moribund. There is no sign whatsoever of heterospory. However, there is some degree of heterothallism. Thus, some spores produce male prothalli, while others produce female ones. The male ones remain male throughout their life, but the female ones eventually change over and become male (effectively, therefore, hermaphrodite). During the change-over period, self-fertilization is possible (Sporne, 1964), for there is no incompatibility mechanism. The peculiar sexual behaviour of *Equisetum*, therefore, confers the benefits of out-breeding during the early phase of prothallus development and the benefits of self-fertilization during the change-over period.

Spores of *Equisetum arvense*, when germinating in a light intensity of 200 foot candles give rise to prothalli, 85 per cent of which are male, but those of *E. telmateia*, germinating in a similar environment, give prothalli only 20 per cent of which are male. Furthermore, under different light intensities, different proportions result, for when germinating in a low light intensity of only 25-foot candles, spores of *E. telmateia* give rise to prothalli 90 per cent of which are male. Another factor affecting the proportion of male prothalli is over-crowding, for in such circumstances only male prothalli result. What the underlying causes may be is still a mystery, but Duckett suggests some biological advantages that might result. Thus, given a wide range of micro-climates within a small area, a wide range of prothallial types in close proximity to each other might result, thereby favouring out-breeding while, at the same time, retaining the potentiality for self-fertilization in a proportion of the prothalli, as a last resort.

The prothallus consists of a flat cushion of tissue, varying in size from 1 mm across to 3 cm in vigorous individuals, The prothalli of such species as *E. arvense*, *E. telmateia* and *E. palustre* have plate-like lamellae projecting upwards, and have projecting antheridia with several cap cells. Those of such species as *E. hyemale* and *E. variegatum* have columnar lamellae (i.e. several cells thick) in which the antheridia are sunken, and have antheridia with two cap cells, looking rather like the guard cells of a stoma.

The archegonia are borne near the base of the lamellae and have projecting necks of two to four tiers of cells in four rows. Instead of a single neck canal cell, there are two boot-shaped cells, lying side by side, as illustrated in Fig. 23H. There is also a ventral canal cell. The antheridia occur round the margin of the basal cushion, but may also occur on the aerial lobes. They are massive and give rise to large numbers of antherozoids, which are spirally coiled and multiflagellate (Fig. 23G).

The first division of the zygote is in a plane more or less at right angles to the axis of the archegonium. No suspensor is formed and the embryo is exoscopic. Fig. 23I shows the spatial relationships of the stem apex (x), the first leaves (1), the root (r) and the foot (f), as described as long ago as 1878, but it is now becoming clear that the various parts of the embryo are not so constant in position and origin as was formerly thought.

There can be little doubt that the Equisetales are related to the Calamitales, but it is most unlikely that they represent their direct descendants. Remains of herbaceous plants resembling *Equisetum* are placed in the genus *Equisetites*. They are traceable right back through the Mesozoic to the Palaeozoic, where several species have been described from the Upper Carboniferous. The situation is thus closely comparable with *Selaginella*, whose herbaceous ancestors were living alongside the related arborescent Lepidodendrales in Carboniferous times.

6 Pteropsida

Sporophyte with roots, stems and spirally arranged leaves (megaphylls) often markedly compound and described as 'fronds' (although some early members showed little distinction between stem and frond). Protostelic, solenostelic or dictyostelic, sometimes polycyclic (rarely polystelic). Some with limited secondary thickening. Sporangia thick- or thin-walled, homosporous or heterosporous, borne terminally on an axis or on the frond, where they may be marginal or superficial on the abaxial surface. Antherozoids multiflagellate.

Some botanists widen the definition of the Pteropsida to include, not only the megaphyllous pteridophytes, but also the gymnosperms and angiosperms, on the supposition that all three groups are related. While this may well be so, it seems preferable to retain the distinction between pteridophytes and seed-plants and to restrict the definition of the Pteropsida so as to exclude all but the ferns. Even so, the group is an enormous one, with over 9000 species, showing such a wide range of form and structure that it is almost impossible to name one character which is diagnostic of the group. The reader will have noticed that almost all of the characters listed at the head of this chapter are qualified in some way.

It will readily be appreciated that, in such a large group, the correct status of the various subdivisions is very largely a matter of personal preference. Accordingly, there are almost as many different ways of classifying the group as there are textbooks dealing with it, and this is particularly true of the fossil members of the group.

At this point, only the major subdivisions are presented, the details being deferred until each subgroup is dealt with.

A   Primofilices*
  1   Cladoxylales*
  2   Coenopteridales*

B   Eusporangiatae
  1   Marattiales
  2   Ophioglossales

C   Osmundidae
    Osmundales

D   Leptosporangiatae
    1   Filicales
    2   Marsileales
    3   Salviniales

PRIMOFILICES

This is a remarkable group of plants that first appeared in the Middle Devonian and survived until the end of the Palaeozoic. As the name suggests, they were probably the ancestors of modern ferns. They may be classified as follows:

1   Cladoxylales*
        Cladoxylaceae*   *Cladoxylon** (*Hierogramma, Syncardia,
                         Clepsydropsis*), *Pseudosporochnus**
                         Calamophyton*

2   Coenopteridales*
        Zygopteridaceae*   *Austroclepsis**, *Metaclepsydropsis**, *Diplolabis**,
                           Dineuron*, *Rhacophyton**, *Ankyropteris**,
                           Etapteris* (= *Zygopteris*, = *Botrychioxylon*)
                           Tubicaulis*, *Protocephalopteris**
        Stauropteridaceae*   *Stauropteris**
        Botryopteridaceae*   *Botryopteris**

The Cladoxylales are a particularly interesting group, whose phylogeny has long been a matter of controversy. On the one hand, they show a number of features in common with the Psilopsida; indeed, not many years have elapsed since *Pseudosporochnus* was transferred from that group (Leclercq and Banks, 1962). On the other hand, they show features in common with the Coenopteridales, whose later representatives had already begun to look fern-like before they became extinct. The group thus stands in an intermediate position which strongly suggests a genuine phylogenetic connection between the two groups.

Several species of *Cladoxylon* are known, of which the earliest is *C. scoparium*, and our knowledge of this is based on one specimen about 20 cm long from Middle Devonian rocks of Germany. According to the reconstruction of the plant by Kraüsel and Weyland (1926) (Fig. 24A), there was a main stem, about 1·5 cm in diameter, which branched rather irregularly. Some of the branches bore fan-shaped leaves (Fig. 24B) ranging in size from 5 mm to 18 mm long.

Some leaves were much more deeply divided than others, but all showed a series of dichotomies. On some of the branches, the leaves were replaced by fertile appendages which were also fan-shaped, each segment terminating in a single sporangium (Fig. 24C).

The vascular system was highly complex and was polystelic; each of the separate steles was deeply flanged; both scalariform and pitted tracheids were present in the xylem. Such complex vascular structure is characteristic of all the species of *Cladoxylon* and some had the additional complication of secondary thickening. *C. radiatum* was similar to *C. scoparium* in that all the xylem was primary, and Fig. 24D illustrates the way in which several xylem flanges were involved in the origin of a branch trace system. It also illustrates the 'islands of parenchyma', as seen in transverse section, which are a common feature of the Zygopteridaceae too. Another feature, shared with the Coenopteridales, is the presence of 'aphlebiae' at the base of the lateral branch (or petiole?). These were similar in position to the stipules of many flowering plants and received separate vascular bundles (1).

Fig. 24E illustrates another type of stem structure, found in *Cladoxylon taeniatum* and several other species, in which each of the xylem strands has an outer region of radially arranged tracheids which are thought to have been formed from a cambium. The arrows in the figure indicate that three of the stem steles were involved in the origin of branch traces. Successive branches, petioles and pinnae of descending order had progressively simpler vascular structures, without secondary wood, and are described under separate form-generic

---

Fig. 24   Cladoxylales

A–H, *Cladoxylon*: A, *C. scoparium*, reconstruction; B, sterile appendage; C, fertile appendage; D, *C. radiatum*, origin of branch traces; E, *C. taeniatum*, t.s. portion of axis near origin of branch trace; F, *Hierogramma*-type stele; G, *Syncardia*-type stele; H, *Clepsydropsis*-type stele. I–L, *Calamophyton*: I, *C. primaevum*, reconstruction; J, t.s. stele; K, *C. bicephalum*, fertile appendage; L, sterile appendage. M–R, *Pseudosporochnus nodosus*: M, reconstruction; N, sterile appendage; O, fertile appendage; P, sporangia; Q, stele of slender axes; R, t.s. stele of stouter axes.
(1, aphlebia traces; 2, peripheral loops of xylem; 3, branch traces.)

(A–C, I, after Kräusel and Weyland; D–H, Bertrand; J, Leclercq and Schweitzer; K, L, Leclercq and Andrews; M–P, Leclercq and Banks; Q, R, Leclercq and Lele.)

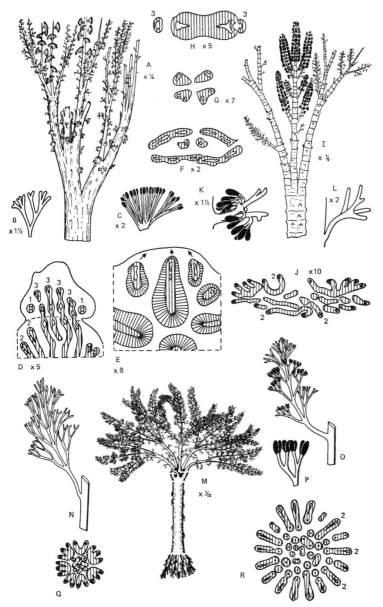

A
x ¼

H x 5

G x 7

F x 2

B
x 1½

C
x 2

K
x 1½

I
x ⅛

L
x 2

D x 5

E
x 8

J x10

N

M
x 1/12

O

P

Q

R

Fig. 24

names. Thus, Fig. 24F shows the *Hierogramma* type of stelar arrange-ment, in which there were six xylem regions, each with islands of parenchyma. Lateral branches from this had four xylem areas and are known as *Syncardia* (Fig. 24G). *Clepsydropsis* (Fig. 24H) was probably the next type of branch, or petiole, although there has been some disagreement among palaeobotanists about this. Its stele, as seen in transverse section, had the shape of an hour-glass (hence the generic name) and from it lateral pinna traces were given off altern-ately, along with a pair of aphlebia traces. It should be noted that similar clepsydroid steles are known from a number of plants belong-ing to the Coenopteridales.

Pseudosporochnus is represented in Middle Devonian rocks from both sides of the Atlantic, but until 1959 our knowledge of the genus was based chiefly on the German species, *P. Krejcii* (Kräusel and Weyland, 1933). It had an erect trunk and a crown of branches which were said to fork dichotomously. However, detailed examina-tion of *P. nodosus* by Leclercq and Banks (1962) and by Leclercq and Lele (1968) has led to a different interpretation. Fig. 24M shows that, instead of forking dichotomously throughout, the aerial branches bore spirally arranged lateral appendages ('fronds') of limited growth. Some were sterile, as in Fig. 24N, while others were fertile, as in Fig. 24O, and bore terminal sporangia (Fig. 24P). There was a trunk, at least 2 m tall and 8 cm thick. The branches were somewhat swollen near the base and had a highly complex stelar anatomy, as illustrated in Fig. 24R. In the centre there were up to 20 centrarch steles and around these were up to 24 radiating mesarch steles, each with a peripheral loop. The vascular anatomy of the more distal regions of the branch system is illustrated in Fig. 24Q. Small traces had their origin from the peripheral loops and supplied the 'fronds'.

Several species of *Calamophyton* are now known from Middle Devonian rocks of Belgium, Germany and the USA. Fig. 24I is taken from an early description of *C. primaevum* by Kräusel and Weyland (1926), but Schweitzer (1973) now believes that this represents merely one branch of a tree with a trunk up to 3 m tall, similar to that of *Pseudosporochnus*. Fig. 24I illustrates a feature which used to be regarded as essential to the definition of the genus, namely, the articulate appearance of the aerial axes. However, other species, e.g. *C. bicephalum*, are not so clearly articulated and, as pointed out by Leclercq and Andrews (1960), it is sometimes difficult to distinguish *Calamophyton* from *Hyenia*. The fertile regions of *C. bicephalum* (Fig. 24K) are certainly very similar to those of

Hyenia (Fig. 21F), differing mainly in the number of sporangia on each fertile appendage. The fertile appendages of *C. primaevum*, however, were said to have been much simpler, forking only once, but it must be remembered that, at that time (1926), the technique of *dégagement* had not been developed; it is possible, therefore, that *C. primaevum* was more complex than was originally thought.

The way in which the fertile appendages of *Calamophyton* were restricted to special fertile branches (Fig. 24I) resembles the early stages of the evolution of a strobilus, and lent support to the idea that *Calamophyton* was an early member of the Sphenopsida. However, it clearly belonged to the Cladoxylales, for its vascular system (Fig. 24J) was composed of irregularly shaped mesarch xylem strands with peripheral loops (Leclercq and Schweitzer, 1965). Whether *Hyenia* also belongs to the Cladoxylales cannot be decided until petrified specimens have been found.

COENOPTERIDALES

This early group of ferns is a large one, consisting of many genera and species. It showed a wide range of growth habit, for some had creeping stems, others had erect trunks and yet others were epiphytes. As with members of the Cladoxylales, here too there is the problem of distinguishing leaf from stem and the term 'phyllophore' is sometimes used for intermediate orders of branching.

Austroclepsis, occurring in Lower Carboniferous rocks of Australia, was first described by Sahni (1928) as a species of *Clepsydropsis*, on account of its clepsydroid petioles. However, the mode of growth of the plant and the internal anatomy of the stem show that it was not a member of the Cladoxylales. It had a stout trunk, at least 30 cm in diameter and 3 m high, and must have looked superficially like modern tree ferns, but it differed fundamentally from these in that, within the mass of roots constituting the main bulk of the trunk, there were several stems instead of just one. These branched within the trunk and gave off numerous petioles in a 2/5 phyllotactic sequence and these, too, continued to run up within the trunk. Each of the many stems had a single stele, usually pentarch, in which there was a central stellate region of mixed pith surrounded by a zone of tracheids. The petioles had a rather narrow clepsydroid stele with two islands of parenchyma bounded by 'peripheral loops' of xylem, and it was from these peripheral loops that pinna traces were given off from alternate sides at distant intervals, each associated with aphlebiae.

Metaclepsydropsis duplex, from Lower Carboniferous rocks of

Pettycur, Scotland, had a creeping dichotomous stem, from which erect 'fronds' arose at intervals (Gordon, 1911). Its stele was circular in cross section or (just before a dichotomy) oval (Fig. 25C), with an inner region of mixed pith and an outer zone of large tracheids. The only protoxylem present was that associated with the origin of a leaf trace, there being no cauline protoxylem at all. The leaf trace was at first oval in cross section (Fig. 25D) but soon became clepsydroid (Fig. 25E). Pinnae were borne in alternate pairs, along with aphlebiae (1). In giving rise to a pair of pinna traces, the peripheral loop (2) became detached and then split into two (3). A new peripheral loop then quickly re-formed (Fig. 25F).

Although the primary pinnae were in alternate pairs, this type of branching did not continue throughout the frond. Secondary and tertiary pinnae were attached singly on alternate sides. The whole structure was three-dimensional and must have looked much more like a branch system than a fern frond.

The reproductive organs of *Metaclepsydropsis* have not been known until very recently, when Chaphekar and Alvin (1972) found a small block of Pettycur Limestone containing fertile material. The sporangia were borne in groups of three or four attached to a common receptacle. Each sporangium was about 1·5 mm long and was broadly banana-shaped. The wall of the sporangium was apparently only one cell thick. Chaphekar and Alvin give the name *Musatea duplex* to this fertile specimen, and they suggest that in future this name should be applied to the whole plant.

Fig. 25   Zygopteridaceae

A, B, *Rhacophyton zygopteroides*: A, reconstruction; B, portion of rachis of sterile frond, showing detail of pinnae. C–F, *Metaclepsydropsis duplex*: C, stem stele; D, petiolar bundle; E, F, rachis bundle, showing origin of pinna traces. G–J, *Etapteris*: G, *E. Scottii*, rachis bundle and origin of pinna traces; H, *E. Lacattei*, reconstruction of sterile frond; I, reconstruction of fertile frond; J, sporangia (stylized).
K–N, *Ankyropteris*: K, *A. Grayii*, stem stele and petiolar bundle; L, *A. westfaliensis*, rachis bundle; M, *Tedelea* (= *Ankyropteris*) *glabra*, reconstruction of pinnules; N, sporangium.
(1, aphlebia traces; 2, peripheral loop of xylem; 3, pinna trace; 4, stem; 5, sterile frond; 6, rachis of fertile fond; 7, pinnae of fertile frond; 8, fertile organs; 9, root.)

(A, B, after Leclercq; C, D, based on Gordon; E, G–I, L, after Hirmer; F, Posthumus; J, Renault and Zeiller; K, based on Scott; M, N, Eggert and Taylor.)

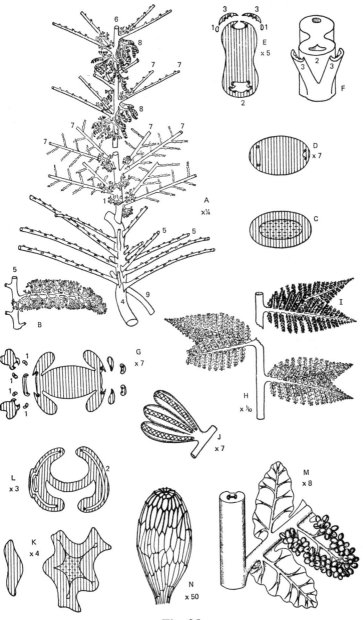

Fig. 25

Diplolabis Roemeri occurs in the same rocks and was very similar to *Metaclepsydropsis*. Its stem anatomy differed, however, in that the inner region of xylem consisted entirely of tracheids. The tracheids of the outer region were arranged in radial rows, but are nevertheless believed by some to have been primary in origin. The petiolar trace had a very narrow 'waist', with the result that it appeared X-shaped in cross section.

Dineuron ellipticum, also from Pettycur, on the other hand had no 'waist' at all in the petiolar trace, which was elliptical in cross section.

In all three of these Lower Carboniferous genera, the origin of pinna traces was the same, suggesting that pairs of lateral pinnae were arranged alternately along the petiole, or phyllophore. The frond was thus a highly compound one whose components formed a three-dimensional structure. However, it had not been realized just how complex they were until the discovery by Leclercq (1951) in the Upper Devonian of Belgium of a specimen of *Rhacophyton zygopteroides* which, although mostly in the form of a compression, nevertheless had some petrified portions. That this plant belonged to the Zygopteridaceae was suggested by the vascular anatomy of the petrified regions, which closely resembled that of the rachis of *Metaclepsydropsis* in being clepsydroid, with peripheral loops. Fig. 25A illustrates a reconstruction of the plant, simplified by the omission of the pinnules. There was a fairly stout stem (4) bearing roots (9) and spirally arranged fronds. The lowermost fronds (5) were sterile and were bipinnate, the ultimate pinnules (Fig. 25B) being dichotomous and apparently without any flattening to form a lamina. The fertile fronds, of which only portions are illustrated (6), were much larger and more complex, for they bore pairs of pinnae (7) arranged alternately, just as had been deduced for *Metaclepsydropsis* from a study of petrified material. Each of the paired pinnae was similar in its mode of branching to a complete sterile frond. Whereas the lower pinna pairs had branched aphlebiae (1), these were replaced higher up by profusely branched structures (8) bearing numerous terminal sporangia. These were about 2 mm long and were without any specially thickened annulus.

Andrews and Phillips (1968) described another species of *Rhacophyton* (*R. ceratangium*) from Upper Devonian rocks of West Virginia, USA which was similar in many respects. Like that of *R. zygopteroides*, the material was partly compressed and partly petrified. Thus, again, both its external form and its internal anatomy are known. Its stems, up to 2 cm in diameter, bore bipinnate vegetative fronds and fertile fronds whose paired pinnae were arranged alternately, as in the Belgian species. The sporangia were borne on profusely dichotomizing, nearly spherical, branch systems attached in pairs in the same relative position as those of *R. zygopteroides*. The most important difference between the two species was that *R. ceratangium* possessed abundant secondary wood. In the light of this fact, Andrews and Phillips suggest that *Rhacophyton* has affinities with the Progymnospermopsida, as a possible fore-runner of the gymnosperms (see Chapter 7). However, the clepsydroid primary stele of the fertile frond rachis and the peculiar branching pattern seem to support Leclercq's contention that the affinities of *Rhacophyton* are, on the contrary, with the Zygopteridaceae.

Schweitzer (1968) has recently described a plant with a similar branching pattern from much earlier deposits, of Middle Devonian age, in Spitzbergen. This plant, which he has called *Protocephalopteris*, is represented by axes which were, presumably, the rachis of a frond on which pairs of pinnae were borne on alternate sides, with aphlebiae at their point of origin. Pinnules were distributed sparsely along the pinnae and, unlike those of *Rhacophyton*, they too were in pairs on alternate sides of the pinnae. Tentatively, Schweitzer suggests that plants such as *Protocephalopteris* and *Rhacophyton* might have evolved from the Lower Devonian *Trimerophyton*.

Etapteris is the name given to petioles in which the 'peripheral loops' remained open throughout (i.e. there were no loops at all). Two pinna traces became detached, fused and then separated again, before passing out into the paired pinnae (Fig. 25G). The Permian species, *E. Lacattei*, is interesting in having progressed further than most other members of the group in the evolution of a photosynthetic lamina, for the ultimate pinnules were flattened (Fig. 25H). In the fertile regions of the frond (Fig. 25I) the pinnules were replaced by groups of sporangia. These were club-shaped, slightly curved, and had a distinct broad annulus of thickened cells (Fig. 25J) Some *Etapteris* fronds were attached to trailing stems, while others belonged to tree-ferns with stout trunks. The nomenclature of the latter is, however, rather troublesome. The names *Zygopteris* and *Botrychioxylon* which have been used are probably synonymous. *Z. primaria* had a trunk about 20 cm in diameter, most of which consisted of a tangle of rootlets and leaf bases. In the centre was a single stem 1·5 cm across, with a five-rayed stele showing the usual two regions, but in this case the outer region looks very much as if it had been formed from a cambium, and many morphologists describe it as secondary wood. *Botrychioxylon paradoxum* had a very similar appearance, but in this stem the cells of the inner cortex were also regularly arranged in radial rows. It would seem, therefore, that the whole of the growing point of the stem must have been organized in a peculiarly regular manner and that great caution should be used in describing even the outer xylem as secondary.

Some species of *Tubicaulis* were tree ferns, while others were epiphytes. They are characterized by a frond form which approached closely to that of a present-day fern, for the fronds coming off in spiral sequence from the stem were pinnate and the pinnae were arranged in one plane.

The same was true of *Ankyropteris*, of which at least eight species are known from Upper Carboniferous deposits both in Europe and in the USA. The genus derives its name from the fact that the petiolar trace in some species, e.g. *A. westfaliensis*, was shaped like a double anchor (Fig. 25L). In some other species the petiolar trace was much less extreme, having less 'waist', but all were alike in that the islands of parenchyma were much extended tangentially and in that the peripheral loop remained closed throughout the origin of a pinna trace. The frond of *A. glabra* (Eggert, 1963) must have looked very much like that of a modern fern for, not only were the pinnae arranged in one plane, but also the pinnules themselves were expanded into a lamina with dichotomizing veins. Eggert and Taylor (1966) described fertile specimens of this fern under the name *Tedelea glabra* (Fig. 25M) and drew attention to the similarity of its sporangia to those of modern members of the Schizaeaceae (cf. Figs. 33 A–D). They were about 0·7 mm long and had an 'annulus' of thick-walled cells occupying almost one-third of the length of the sporangium. Because of its expanded pinnules and the structure of its sporangia, Eggert and Taylor have suggested that *Tedelea (Ankyropteris) glabra* should be transferred from the Zygopteridaceae to the Filicales. The presence of a peripheral loop in the xylem of the rachis and the presence of aphlebiae at the point of attachment of the pinnae, however, must not be overlooked, for they are both characteristic features of the Zygopteridaceae. Perhaps the most interesting conclusion is that in *Tedelea* we are seeing signs of an evolutionary relationship between the Coenopteridales and the Filicales.

Ankyropteris Grayi, from British coal measures, had a stem of considerable length which was over 2 cm in diameter. It was probably a climbing plant. The petioles were borne in a 2/5 phyllotactic spiral, corresponding to the five rays of the stellate stele (Fig. 25K). As in other members of the group, there were two distinct regions in the stele, viz. an inner region composed of a mixture of tracheids and parenchyma, and an outer one entirely of tracheids. However, there is no suggestion that the outer region might have been formed from a cambium, for the cells within it were not arranged in radial rows.

Stauropteris is represented by two species, *S. burntislandica* from the Lower Carboniferous and *S. oldhamia* from the Upper Carboniferous. Although the method of branching of the frond was similar to that of many of the Zygopteridales, differences in the vascular system are sufficient to warrant the creation of a separate family. The most important of these is the absence of islands of parenchyma in the xylem of the petiolar traces. It is believed that, so far, only portions of fronds have been found and that the stems have yet to be discovered. Fig. 26A shows how the frond of *S. burntislandica* was, constructed, pairs of pinnae arising alternately along the petiole each associated with aphlebiae. Then each pinna gave rise to secondary pinnae in the same way and this pattern was repeated at all levels of branching within the frond. In this respect, *Stauropteris* resembled *Protocephalopteris*, to which Schweitzer (1968) believes it may be related. Starting from the Lower Devonian *Trimerophyton*, he suggests that there were two divergent lines of evolution, of which one led through plants like *Rhacophyton* to *Etapteris*, while the other led through plants like *Protocephalopteris* to *Stauropteris*. The vascular system of the petiole of *S. oldhamia* (Fig. 26B) consisted of four regions of xylem either contiguous or separate from each other, each with a mesarch protoxylem. The smaller branches, however, tended to have a single tetrarch strand.

Perhaps the most interesting feature of all about *Stauropteris burntislandica* is the fact that it was heterosporous. Its megasporangia (Fig. 26C), when found isolated, are called *Bensonites fusiformis*. They were strangely fleshy at the base and most commonly contained two functional megaspores along with two very small and, presumably, abortive ones, although examples have been found with four, six or eight megaspores. It is believed that the whole structure was shed from the parent plant without prior dehiscence. The microsporangia of *S. oldhamia* (Fig. 26D) were spherical, were typically eusporangiate in having a thick wall and had a terminal stomium where dehiscence took place, but there was no annulus of thick-walled cells.

In the past, the Botryopteridaceae were often described as much simpler in their organization than the rest of the Coenopteridales, but recent investigations on both sides of the Atlantic have demonstrated that this is far from the case. *Botryopteris antiqua*, from the Lower Carboniferous of Scotland, is the earliest known species and was also the simplest in its internal anatomy. It had trailing dorsi-

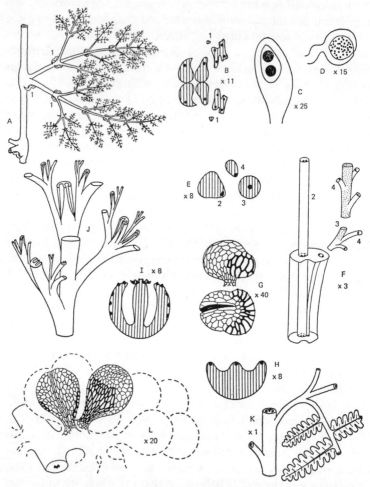

Fig. 26   Stauropteridaceae and Botryopteridaceae

A–D, *Stauropteris*: A, *S. burntislandica*, reconstruction of part of frond;
B, vascular system of *S. oldhamia*; C, *Bensonites fusiformis*
(megasporangium of *S. burntislandica*); D, microsporangium. E–L,
Botryopteris: E, vascular strands of *B. antiqua* in t.s.; F, the same, as a
reconstuction; G, sporangia of *B. antiqua*; H, petiolar vascular strand
of *B. ramosa*; I, petiolar strand of *B. forensis*; J, vascular system of
B. trisecta; K, reconstruction of part of frond of an advanced species,
with flattened pinnules; L, sporangia of *B. forensis*.

(A, C, after Surange; B, Bertrand; D, Scott; F, G, L, after Galtier;
I, Corsin; J, Andrews; K, Delevoryas and Morgan.

ventral axes up to 2 mm in diameter tentatively described as petioles on which were borne erect or semi-erect radial stems, which in turn bore further petioles, in spiral succession, together with roots. The steles of these various axes are illustrated diagrammatically in Fig. 26E, where '2' indicates the one belonging to the trailing dorsiventral petiole. It was a solid rod of tracheids with multiseriate pits or with scalariform or reticulate pits. One or two protoxylem regions occurred in an almost exarch position. The radial stems '3' were about the same diameter, but the stele was circular in cross section with the smallest tracheids (protoxylem?) in the centre. The petioles that were borne on these stems were somewhat smaller, up to 1·4 mm in diameter, and had an oval stele with a lateral protoxylem, '4'. They underwent branching, of up to five successive orders, to produce a multipinnate branch system, the stele becoming smaller and smaller until the ultimate pinnules had only a few tracheids or even only one. There was no flattening of the pinnules anywhere in the frond of this species and the distinction between stem and petiole is purely arbitrary. The manner in which 'stems' were borne on 'petioles' finds a parallel at the present day in several members of the Filicales, which are described as bearing 'petiolar shoots' (Troop and Mickel, 1968). First described by Surange (1952) for *Botryopteris antiqua* in calcified specimens from Scotland, this arrangement has since been confirmed in silicified specimens from France by Galtier (1969), whose reconstruction is illustrated in Fig. 26F. The sporangia of this species were globose, up to 0·25 mm across, and had a multicellular annulus placed laterally (Fig. 26G).

A comparison of this early species with those of the Upper Carboniferous and the Permian shows that there was a trend in the evolution of the petiole trace towards a greater degree of dorsiventrality, together with an increase in the number of protoxylems. Thus, *Botryopteris ramosa* (Upper Carboniferous) had a shallow gutter-shaped stele with three protoxylems (Fig. 26H), whereas *B. forensis* (Permian) had a stele shaped like the Greek letter ω in transverse section, with up to fifteen protoxylems (Fig. 26I). Some of these later species, furthermore, are known to have had laminate pinnules (Fig. 26K).

The complexity of the branching of the later species of *Botryopteris* is illustrated by the reconstruction of the stelar system of *B. trisecta* (Fig. 26J). Its erect stem had a cylindrical protostele and bore leaves in a spiral sequence. The petioles had an oval vascular strand and branched into three. The two lateral branches then trisected again but, whereas the median traces in each case were ω shaped, the lateral ones were cylindrical, like the stem stele. The whole frond was arranged in three dimensions, except for the ultimate pinnules which were disposed in one plane.

Associated with this plant were found some remarkable spherical masses containing thousands of sporangia, which are believed to represent the fertile parts of the frond, although in the meantime they are described under a separate specific name, *Botryopteris globosa*. The whole mass was up to 5 cm across and had, running through it, a system of branches with ω shaped steles. *B. forensis* (Fig. 26L) was essentially the same, as shown by Galtier (1971). Each sporangium was pear-shaped, with an annulus of thick-walled cells on one side and a line of elongated thin-walled cells forming an annulus running up the other side. In most species of *Botryopteris*, the sporangium wall is described as only one cell thick, suggesting that, in this respect at least, they were leptosporangiate. It is apparently true of some of the sporangia of *B. globosa*, but not of all, for some clearly had a second layer of thin-walled cells on the inside. This may well have shrivelled after the spores had been shed, so becoming invisible when petrified. Thus, although approaching the leptosporangiate condition, *B. globosa* had certainly not yet achieved it, and the same is probably true of all the species.

Eusporangiatae
  Marattiales
    Asterothecaceae*   *Psaronius*, Asterotheca*, Scolecopteris*,*
                       Acitheca*,   Eoangiopteris*
    Angiopteridaceae *Angiopteris*
    Marattiaceae *Marattia*
    Danaeaceae *Danaea*
    Christenseniaceae *Christensenia*
  Ophioglossales
    Ophioglossaceae *Ophioglossum, Botrychium, Helminthostachys*

MARATTIALES

It was customary in the past to describe the Carboniferous as the Age of Ferns. This was because of the abundance of large fern-like fronds in the coal measures, but it is now known that many of them really belonged to gymnosperms, for they have been found in association with seeds. Indeed, it is now suspected that most of them were gymnospermous. However, there can be no certainty about sterile fronds and these must, therefore, be placed in a number of form genera defined on the basis of the overall shape of the frond and on the shape and venation of the pinnules. *Pecopteris* is one of these and a large number of species are known. Some of them were certainly gymnosperms, but others were equally certainly ferns, for they bore sori of thick-walled sporangia. The frond, sometimes as much as 3 m long, was many times pinnate and the pinnules were attached along their entire base, each with a single midrib. The lateral veins were somewhat sparse and branched dichotomously once or twice (Figs. 27A and 27D) or remained unbranched.

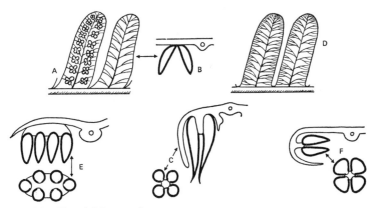

Fig. 27   *Fossil Marattiales*

A, B, *Asterotheca*: A, fertile pinnules; B, sorus. C, D, *Acitheca*: C, sorus; D, sterile pinnules. E, *Eoangiopteris*, sorus. F, *Scolecopteris*, sorus.

(A, after Hirmer; B, E, F, Mamay; C, Renault; D, Brogniart.)

Asterotheca is the name given to pecopterid fronds bearing sessile sori made up of four or five sporangia fused at the base into a synangium, but with the distal part free (Fig. 27B). The sori were commonly arranged in two series along the pinna, as illustrated in Fig. 27A, each associated with a veinlet in the lamina. *Scolecopteris* was similar, except that the sorus was elevated on a short pedicel, or receptacle (Fig. 27F). In *Acitheca*, the sporangia were elongated and pointed and were arranged round a central plug-like receptacle (Fig. 27C). *Eoangiopteris* is regarded by Mamay (1950) as a more advanced type of sorus since it was linear instead of radial. Each had a cushion-like receptacle, on which were five to eight sporangia (Fig. 27E). In all these genera, the sporangium wall was very massive, many cells thick, and the number of spores liberated from each was very high, e.g. up to 2000 in *Asterotheca parallela*.

Fronds of these various types are often found in association with stout trunks that bore a superficial resemblance to those of modern tree-ferns, some of them as much as 15 m high, but organic connection between fertile frond and trunk has only been demonstrated conclusively on two occasions (Stidd, 1971). The evidence, nevertheless, suggests that *Asterotheca* fronds were borne in a crown at the summit of trunks known as *Psaronius*. Many species, belonging to a number of subgenera of *Psaronius*, are known and most of them had remarkably complex stelar anatomy. Some of them were as much as 75 cm across, but most of this width was occupied by a thick mantle of roots, for the single stem in the centre was only a few cm in diameter. The stele of most species was a polycyclic dictyostele which, in the more complex types, contained as many as eleven interconnecting coaxial cylinders (or, rather, inverted cones fitting inside one another). Each was dissected into a number of mesarch meristeles completely surrounded by phloem, and the leaf traces at any particular level arose from the outermost system, while the inner systems were concerned with the origin of leaf traces at higher levels. The earliest examples, however, were simpler than this in their internal anatomy, e.g. *P. Renaultii* from the Lower Coal Measures had an endarch solenostele; and there is evidence that even the complex Permian species had a relatively simple structure near the base of the trunk, as would be expected by analogy with present-day ferns (Morgan, 1959). Although the trunks were widest at the base, this was not because the stem within was wider but because there was a greater number of rootlets in the mantle; the stem was actually smaller towards the base. Some species had the leaves arranged distichously, some in three or four vertical rows, while others had them arranged spirally, as in most modern representatives of the group.

The Marattiales are represented at the present day by about 200 species, placed in six (or seven) genera, most of which are confined to the tropics. *Angiopteris* (100 species) is a genus of the Old World, extending from Polynesia to Madagascar, while *Danaea* (thirty-two species) is confined to the New World. *Marattia* (sixty species) is pan-tropical and extends as far south as New Zealand. *Christensenia* (= *Kaulfussia*) is monotypic and is confined to the Indo-Malayan region. Most species have massive erect axes, but they never attain the dimensions of the fossil *Psaronius*. The largest, although reaching a diameter of 1 m, seldom exceed this in height. *Christensenia* and some species of *Danaea*, however, have creeping horizontal axes. The fronds of some species are larger than in any other living ferns and may be as much as 6 m long, with petioles 6 cm in diameter.

They may be as much as five times pinnately compound or, in some species, only once pinnate, like a Cycad leaf, while a few species have a simple broad lamina. *Christensenia* is peculiar in having a palmately compound frond, as the specific name, *C. aesculifolia*, implies. It is also peculiar in having reticulate venation, for all the other genera have open dichotomous venation. All show circinate vernation, i.e. the young frond is coiled like a crozier and gradually uncoils as it grows. This is a feature which they share with Leptosporangiate ferns but which is absent from the Ophioglossales. With the exception of *Danaea trichomanoides*, all the living members of the group have very leathery pinnules in whose ontogeny several rows of marginal initials are active (instead of a single row of marginal initial cells, as is more usual in leaves of other plants). In many species there are swellings, or pulvini, at the base of the pinnae and pinnules, which play a part in the geotropic responses of the leaf, and in all species there are thick fleshy stipular flanges at the base of the petiole. Fig. 28J illustrates the appearance of the growing point of *Christensenia*, showing how the stipules (1) are joined by a commissure (2), and how they are folded over the primordium. After the frond has died and has been shed, the stipules and the leaf base remain attached to the axis and contribute much to its overall diameter.

The young parts of *Marattia* and *Angiopteris* are covered with short simple hairs, while those of *Christensenia* and *Danaea* bear peltate scales. Bower (1923) suggested that the nature of the dermal appendages in ferns can be a useful indicator of primitiveness or advancement, hairs being more primitive than scales; on this basis therefore, *Christensenia* is relatively advanced and this conclusion is supported by its possessing reticulate venation. A comparison of fossil and recent members of the group suggests that there has been progressive reduction in height, from the tree-like Carboniferous forms, through an intermediate stumpy erect axis, to an oblique or horizontal creeping rhizome; and on this basis, too, *Christensenia* along with *Danaea* is to be regarded as relatively advanced.

The stem grows by means of a bulky type of meristem, not referable to a single initial cell (Bower, 1926) and is characterized by the absence of sclerenchyma. Mucilage canals and tannin cells are abundant throughout and give the tissues a very sappy texture. The vascular anatomy of the stem is the most complex of all living pteridophytes and is surpassed in complexity only by fossil members of the group, such as *Psaronius*. A transverse section of the stem of *Angiopteris* (Fig. 28N) reveals a number of concentric rings of meristeles which, in a dissection (Fig. 28M), are seen to be part of a

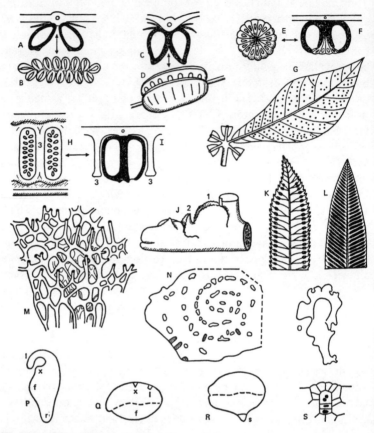

Fig. 28   Marattiales

A, B, sorus of *Angiopteris*. C, D, sorus of *Marattia*. E, F, sorus and G, pinna of *Christensenia*. H, I, sorus of *Danaea*. J, rhizome and leaf base of *Christensenia*. K, pinna of *Marattia*. L, pinna of *Danaea*.

M, N, vascular system of *Angiopteris*: M, dissected; N, as seen in t.s.

O, prothallus of *Marattia*. P, Q, old and young embryos of *Angiopteris*. R, embryo of *Danaea*. S, archegonium of *Marattia*.

(1, stipule; 2, stipular flange; 3, flange, or partition, on lamina; f, foot; 1, leaf; r, root; s, suspensor; x, stem apex.)

(A–I, K, L, after Bower; J, Gwynne-Vaughan; M, N, Shove; O, Q–S, Campbell; P, Farmer.)

series of complex and irregular meshworks lying one within the other, yet interconnected by 'reparatory strands'. The whole system may be described as a highly dissected polycyclic dictyostele, but can best be visualized as a series of inverted cones of lace stacked inside each other. Although each meristele in the sporeling is surrounded by an endodermis, in the adult state the endodermis is completely lacking.

The earliest protoxylem elements to lignify are 'annular-reticulate', i.e. adjacent rings of lignin are interconnected by a network of strands, whereas later ones are reticulate. The metaxylem elements are scalariform and, in *Angiopteris*, the orientation of the elongated bordered pits is sometimes longitudinal, instead of transverse. This peculiar arrangement has been called 'ob-scalariform' and occurs elsewhere in the Ophioglossaceae and a few leptosporangiate ferns (*Dennstaedtia* and *Blechnum*) (Bierhorst, 1960).

Each leaf, in a mature plant, receives a number of traces which arise from the outermost system of meristeles (the cut ends of the leaf traces are shown in solid black in Fig. 28M), but the root traces may arise from the innermost regions of the stele, threading their way through successive cones on their way to the cortex (cross-hatched in Fig. 28N). In those species with erect axes, the roots may emerge from the cortex some distance above the ground, so forming prop-roots. They are polyarch, with as many as nineteen exarch protoxylems and, while the aerial portions are medullated, as soon as the roots penetrate the soil the xylem extends right to the centre. Those of young plants usually contain a mycorrhizal fungus within the cortex (an oomycete known as *Stigeosporium marattiacearum*).

In all genera, the sori are borne in a 'superficial' manner, i.e. on the dorsal surface of the lamina, and beneath a vein or a veinlet. *Christensenia* has circular sori irregularly distributed between the main veins (Fig. 28G) but, in all other genera, the sorus is more or less elongated beneath a lateral vein (Figs. 28K and 28L). In *Angiopteris* the sporangia are free from each other (Figs. 28A and B), but in *Marattia*, *Danaea* and *Christensenia* they are fused into a synangium (Figs. 28C, D, E, F, H and I). *Danaea* is peculiar in having fleshy flanges of tissue (3) projecting between the adjacent synangia (or, according to some, in having the synangia sunken into a very fleshy pinnule).

The first stage in the development of a sporangium is a periclinal division of a single epidermal cell, of which the inner half gives rise ultimately to the archesporial tissue, while the outer half gives rise to part of the sporangium wall, the rest of the wall being produced by the activity of adjacent cells. At maturity, the sporangium wall is many cells thick and there is a tapetum formed from the innermost wall cells. The occurrence of numerous stomata in the sporangium wall is an interesting feature rarely found elsewhere and presumably associated with its massive structure. Very large numbers of spores are produced from each sporangium (e.g. 1440 in *Angiopteris*, 2500 in *Marattia* and over 7000 in *Christensenia*) and, since all the sporangia within a sorus mature and dehisce simultaneously, prodigious numbers of spores are shed.

In those species with free sporangia, e.g. *Angiopteris*, there is a crude kind of annulus of thickened cells, whose contractions pull the sides of the sporangium apart along a line of dehiscence on the inner face (Fig. 28B). Those with synangia have no such device; instead, a thin part of the sporangium wall dries and shrinks to form a pore through which the spores can fall (Figs. 28E, F, H, I). The whole sorus in *Marattia* is very woody and, when ripe, splits into two halves which are slowly pulled apart, so as to expose the pores in each sporangium (Fig. 28D).

Germination of the spores is rapid, occurring within a few days of being shed, and they develop directly into a massive dark green thalloid prothallus, which is mycorrhizal and is capable of living for several years. An old prothallus may be several centimetres long and may resemble closely a large thalloid liverwort (Fig. 28O). The prothallus is monoecious but, while the antheridia occur on both the upper and lower surface, the archegonia are confined to the lower surface, where they occur on the central cushion along with rhizoids. Both types of gametangia are sunken beneath the surface of the prothallus and the antheridium is large and massive. The archegonium (Fig. 28S) has a large ventral canal cell (except in *Danaea*) and a neck canal cell with two nuclei. The antherozoids are coiled and multi-flagellate, as in other ferns.

The first division of the zygote is at right angles to the axis of the archegonium, and the embryo is endoscopic. Thus, since the archegonial neck is directed downwards, the embryo is orientated with its shoot uppermost and, as it grows upwards, it bursts its way through the tissues of the prothallus. A minute suspensor is present in *Danaea* (Fig. 28R) and in some species of *Angiopteris*, but *Marattia*, *Christensenia* and most species of *Angiopteris* are completely without a suspensor. This lack of constancy is paralleled in the Ophioglossales and has led to speculation as to its phylogenetic implications. A suspensor is generally held to be a primitive character and its presence even if not universal in the Eusporangiatae, places them at a lower level of evolution than the remaining ferns, from which it is completely absent.

The epibasal hemisphere gives rise to the shoot apex (x) and the first leaf (1) (Fig. 28Q), but there is no regular pattern of cell division and the hypobasal region gives rise to a poorly developed foot (f) and, somewhat later, to the first root (r) (Fig. 28P).

Chromosome counts give a haploid number $n = 40$ in *Angiopteris*.

OPHIOGLOSSALES

This group of plants, completely without any early fossil record, is represented by about eighty living species, belonging to three genera. *Botrychium* (thirty-five species) is cosmopolitan in distribution and *Ophioglossum* (forty-five species) is nearly so, but *Helminthostachys* (monotypic) is restricted to Indo-Malaysia and Polynesia. Two species are fairly common in the British Isles, *Botrychium lunaria*, 'Moonwort' (Fig. 29A) which grows in dry grassland and on rocky ledges, and *Ophioglossum vulgatum*, 'Adder's Tongue' (Fig. 29H) in damp grassland, fens and dune-slacks, while a third species, *O. lusitanicum*, is restricted to grassy cliff tops in the Channel Islands and the Scilly Isles.

The stem, in most species, is very short and is erect, except in a few epiphytic species of *Ophioglossum* and in *Helminthostachys*, where it becomes a horizontal rhizome as the plant grows larger. Where the stem is erect, the leaves arise in a spiral sequence, but in temperate regions it is normal for only one leaf to be produced each year. In *Helminthostachys*, the leaves are borne in two ranks along the rhizome; they are large and ternately compound, but in the other two genera they are usually much smaller. Those of *Botrychium* are pinnately compound; those of *Ophioglossum* are simple or lobed and, unlike those of the other two genera, have a reticulate venation. At the base of the petiole there is a pair of thin stipules which enclose the apical bud; and the next leaf, when it begins to grow, has to break its way through the thin sheath covering it. Unlike all other living ferns their leaves are not circinately coiled when young.

In all three genera, the fertile fronds have two distinct parts, the fertile part being in the form of a spike which arises at the junction of the petiole with the sterile lamina, on its adaxial side. The fertile spike is pinnately compound in those genera with a compound lamina and simple in *Ophioglossum*, where the lamina is simple. Its morphological nature has been the subject of some considerable discussion in the past but is now generally thought to represent two basal pinnae which have become ontogenetically fused, face to face (i.e. it is believed that some early ancestor of the group had two fertile basal pinnae, whose primordia became fused during subsequent evolution). Today, the only evidence for the double nature of the spike lies in its vascular supply.

The roots are peculiar in being completely without root hairs, a feature which is possibly connected with their mycorrhizal habit.

Growth of the stem apex is from a single apical cell, and its products are characteristically soft and fleshy, for they are without sclerenchyma. The stem of the young sporeling is protostelic, but soon becomes medullated. Later on, the stem of *Botrychium* becomes solenoxylic, i.e. there are leaf gaps in the xylem, but not in the single external endodermis. Ultimately, the appearance of a sporadic internal endodermis may give rise to a rudimentary solenostele. *Botrychium* is the only genus of living ferns to show secondary cambial activity, and in some species it may give rise to a considerable thickness of secondary wood, composed of tracheids and woodrays. Rhizomes of *Helminthostachys* pass through much the same stages of stelar organization, but the largest specimens go one stage further and achieve true solenostely, with an internal as well as an external endodermis.

Ophioglossum varies considerably in its internal anatomy, according to species. Some possess an outer endodermis, but in most species it is absent, even in the young stages. The leaf gaps in the xylem overlap one another, giving rise to a network of meristeles, which form a rudimentary kind of dictyostele. The xylem is endarch in *Botrychium* and *Ophioglossum*, but mesarch in *Helminthostachys*. The earliest formed protoxylem tracheids are very similar to those of the Marattiales; later ones are reticulate (some being ob-reticulate) but scalariform tracheids are absent (Bierhorst, 1960). A pronounced feature of all three genera is the distinctly bordered circular pits in the metaxylem tracheids, but early accounts of the universal presence of a torus in the pit closing membrane appear to be incorrect. Bierhorst records them only in *Botrychium dissectum* and states that even in this species they are not a constant feature.

The sporangia in all three genera are 'marginal' in origin. In *Botrychium*, they are borne in two rows along the ultimate pinnules of the fertile spike (Fig. 29B) and each receives its own separate vascular supply from a vein running into the pinnule (Fig. 29C). In *Helminthostachys*, the axis of the fertile spike bears numerous 'sporangiophores' in several rows, each bearing several sporangia and a few tiny green lobes at the tip. The spike of *Ophioglossum* bears two rows of sporangia fused together, beyond which the axis projects as a sterile process (Fig. 29H). A number of vascular bundles run longitudinally up the middle, anastomosing occasionally and giving off lateral branches to the sporangia (Fig. 29J).

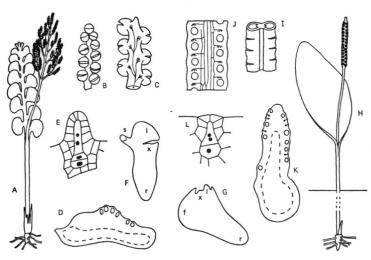

Fig. 29   Ophioglossales

A–G, *Botrychium*: A, *B. lunaria*; B, fertile pinnule; C, vascular supply to sporangia; D, prothallus of *B. virginianum*; E, archegonium; F, embryo of *B. obliquum*; G. embryo of *B. lunaria*; H–L, *Ophioglossum*: H, *O. vulgatum*; I, portion of fertile spike; J, vascular supply to sporangia; K, prothallus; L, archegonium of *O. pendulum*. (f, foot; l, leaf; r, root; s, suspensor; x, stem apex)

(A, H, after Luerssen; B, I, Bitter; C, Goebel; D, L, Campbell; E, Jeffrey; F, Lyon; G, K. Bruchmann.)

Early stages of development of the sporangium are similar to those in the Marattiales; a single initial cell undergoes a periclinal division, the inner half giving rise ultimately to the archesporial tissue, while the outer half goes to form part of the sporangium wall. Adjacent cells contribute further to the wall, which is very massive and several cells thick at maturity. A tapetum of several layers of cells is formed from the inner regions of the sporangium wall, which break down to form a continuous plasmodium in which the spores develop. As in Marattiales, there are stomata in the sporangium wall.

Dehiscence of the sporangium is transverse in *Botrychium* and *Ophioglossum* (Figs. 29B and I), but longitudinal in *Helminthostachys*, and large numbers of spores are released (more than 2000 in *Botrychium* and as many as 15 000 in *Ophioglossum*).

The prothallus in all three genera is mycorrhizal. Indeed, the presence of the appropriate fungus is essential for the growth of the prothallus beyond the first few cell divisions. In most cases the prothallus is deeply buried in the soil and lacks chlorophyll, but cases have been reported of superficial prothalli, in which some chlorophyll was present. Some have abundant rhizoids, but others are completely without them.

The prothallus of *Botrychium virginianum* (Fig. 29D) is a flattened tuberous body, up to 2 cm long. Antheridia appear first and are deeply sunken. Large numbers of antherozoids are liberated from each and escape by the rupturing of a single opercular cell. The archegonium has a projecting neck several cells long, a neck canal cell with two nuclei, and a ventral canal cell (Fig. 29E).

The prothallus of *Ophioglossum vulgatum* differs in being cylindrical, and may be as much as 6 cm long (Fig. 29K). Frequently, there is an enlarged bulbous base, in which the bulk of the mycorrhizal fungus is located. (In both Figs. 29D and K, the extent of the fungus is indicated by a broken line.) As in *Botrychium*, the antheridia are sunken and produce very large numbers of antherozoids. Unlike *Botrychium*, however, its archegonia are sunken too. In Fig. 29L, the archegonium of *O. pendulum* is illustrated at a stage just before maturity, when there are visible two nuclei in the neck canal cell, but just before the basal cell has divided. Indeed, a ventral canal cell has rarely been seen, presumably because it disintegrates almost as soon as it is formed.

As in the Marattiales, the first division of the zygote is in a plane at right angles to the archegonial axis. In *Helminthostachys*, the outer (epibasal) hemisphere undergoes a second division, so as to produce a suspensor of two cells, while the hypobasal hemisphere gives rise to a foot, a root and, later, the stem apex. The embryo is thus endoscopic, but during its further development its axis becomes bent round through two right angles, so as to allow the stem to grow vertically upwards. The embryo of some species of *Botrychium* is likewise endoscopic and has a small suspensor (Fig. 29F), but in others including *B. lunaria* (Fig. 29G), there is no suspensor and the embryo is exoscopic; and this is true of all species of *Ophioglossum*. In all cases there is considerable delay in the formation of the stem apex, and in some species it may be several years before the first leaf appears above the ground, by which time many roots may have been formed. These long delays suggest that the mycorrhizal association is an important factor in relation to the nutrition, not only of the prothallus, but also of the young sporophyte.

Chromosome counts show a surprising range within the group, for *Botrychium* has a haploid number n = 45, *Helminthostachys* n = 46 or 47, while in *Ophioglossum vulgatum* n = 250–260 and in *Ophioglossum reticulatum* n = 631 + 10 fragments.

Despite these divergent chromosome numbers, there can be little doubt that the three genera of the Ophioglossales are fairly closely related, nor that they represent an ancient and primitive group of ferns despite the lack of fossil representatives. The reticulate venation of *Ophioglossum*, its consolidated fertile spike and its complete lack of a suspensor together suggest that it has reached a more advanced stage of evolution than either of the other two genera. As in the Marattiales, it seems that the upright stem is the basic condition, since even in *Helminthostachys* the young plant has an erect axis.

Regarding the relationships between the Ophioglossales and the Marattiales, it is not easy to decide which characters are significant. Of the many characters common to the two groups, most indicate merely that they have reached roughly the same stage of evolution, rather than that they are closely related. These may be briefly listed as 1. basically erect axis, 2. stipules at the base of the petiole, 3. absence of sclerenchyma, 4. sporadic endodermis, 5. massive sporangium wall, with stomata, the sporangia showing a tendency to fusion, 6. large spore output, 7. prothallus long lived, 8. massive antheridium, 9. suspensor present in some, absent in others. Characters which suggest that the two groups are only distantly related are the circinate vernation of the Marattiales and their superficial sori, contrasting with the absence of circinate vernation from the Ophioglossales and their marginal sporangia.

OSMUNDIDAE

Osmundales
    Osmundaceae   *Zalesskya*, Thamnopteris*, Osmundites*,*
                      Osmunda, Todea, Leptopteris

The modern representatives of the Osmundales occupy an isolated position among the ferns, intermediate in many respects between the Eusporangiatae and the Leptosporangiatae but not necessarily, therefore, linking the two groups phylogenetically, for they are an extremely ancient group with an almost complete fossil history extending as far back as the Permian. Those that have survived to the present day can truly be described as 'living fossils'.

E

All have erect axes, bearing a crown of leaves; and the same is true of the fossil members, some of which had trunks 1 m or more in height. Among the earliest representatives, in the Permian, were several species of *Zalesskya*. These had a solid protostele in which there were two distinct regions of xylem (an inner region of short tracheids and an outer one of elongated tracheids forming an unbroken ring). The same was true of *Thamnopteris Schlechtendalii*, but *T. Kidstonii* had a slightly more advanced stelar anatomy, in that the central region was occupied by a mixed pith of tracheids and parenchyma. *Osmundites Dunlopii* from the Jurassic was similar to *T. Kidstonii*, but the contemporaneous *O. Gibbeana* showed some dissection of the xylem ring into about twenty separate strands (Kidston and Gwynne-Vaughan, 1907). Nevertheless, the stele was still strictly a protostele, since there was a continuous zone of phloem (and, presumably, endodermis) round the outside. The term 'dictyoxylic stele' can conveniently be used to describe this arrangement. Poor preservation does not allow any statement to be made about the central pith regions of these two forms, but in the Lower Cretaceous *O. Kolbei* there was definitely a mixed pith. The Cretaceous species *O. skidegatensis* had a pith of pure parenchyma and showed a further advance in having some internal phloem, while *O. Carnieri* was the most advanced of all, in being truly dictyostelic. This is most interesting, for it is a condition not achieved by any modern representatives of the group. Most of these are no further advanced in stelar anatomy than the Jurassic *Osmundites Gibbeana*.

Of the living genera, *Osmunda* (fourteen species) is widespread in both hemispheres, *Leptopteris* (six species) is confined to Australasia and the South Sea Islands, while *Todea* is represented by the single species *T. barbara*, found in S. Africa and Australasia. (Some taxonomists include *Leptopteris* in the genus *Todea*.) Only one species, *Osmunda regalis* – the 'Royal fern' – is represented in the British flora. Its stems are massive and branch dichotomously to form large hummocks. *Todea barbara* may have a free-standing trunk 1 m or more high, and so also may *Leptopteris hymenophylloides*, while one species of *Leptopteris* from New Caledonia attains a height of 3 m.

A transverse section of the stem of a mature *Todea* (Fig. 30K) exhibits a typical dictyoxylic condition. The central medulla is surrounded by separate blocks of xylem, outside which there is phloem and a continuous endodermis. Occasionally, some internal phloem occurs, but no internal endodermis. Most species of *Osmunda* are similar, but *O. cinnamomea* sometimes has an internal, as well as external, endodermis (Fig. 30B). The types of xylem element present are similar, in some respects, to those of the Marattiaceae (Bierhorst, 1960) and the position of the protoxylem ranges from endarch in *Todea* to nearly exarch in *Osmunda*.

The leaves, in most species, are leathery in texture, but those of *Leptopteris hymenophylloides* are comparable with those of the Hymenophyllaceae ('filmy ferns') and have a thin pellucid lamina, only two or three cells thick, from which stomata are completely

Fig. 30   Osmundales

A–F, *Osmunda*: A, partly fertile frond of *O. regalis*; B, stele of *O. cinnamomea*; C, prothallus of *O. Claytoniana*; D, antheridium; E, archegonium; F, sporangium of *O. japonica*. G, *Leptopteris hymenophylloides*, fertile pinnule. H–K, *Todea barbara*; H, fertile pinna; I, J, sporangial primordia (I with tetrahedral archesporial cell, J with cubical archesporial cell); K, stele.

(A, F, H, after Diels; C–E, Campbell; G, Hewitson; I, J, Bower; K, Seward and Ford.)

lacking. During their development the leaves of all species exhibit circinate vernation and are covered with hairs. The base of the petiole is broad and winged in a manner reminiscent of the Eusporangiatae and, after the frond has been shed, the leaf base is persistent, adding considerably to the diameter and the mechanical strength of the stem.

The fronds of *Osmunda regalis* are twice pinnate, those produced first in each season being sterile. These are followed by partially fertile fronds (Fig. 30A), while the last to be produced are often completely fertile. The fertile pinnules are very reduced tassel-like structures, representing just the midrib. In the absence of a lamina, the sporangia cannot be 'superficial' and are usually described as 'marginal'. In partially fertile fronds of *O. regalis*, the fertile pinnules occupy the distal regions, but in those of *O. Claytoniana* they occupy the middle regions. *Todea barbara* has twice-pinnate fronds in which the fertile pinnules show scarcely any modification and the sporangia are superficial, being densely scattered over the under-surface of the lamina. They occupy the basal regions of partially fertile pinnae (Fig. 30H). The fronds of *Leptopteris hymenophylloides* are large and feathery, with deeply divided pinnules. The sporangia are scattered sparsely along the veinlets of unmodified pinnules (Fig 30G). In no case is there any tendency for the sporangia to become aggregated into sori, nor is there any sign of an indusium.

The sporangium is not strictly leptosporangiate, for several cells play a part in its initiation and, at maturity, it is relatively large and massive with a stout short stalk. There is some variation in the shape of the archesporial cell, as illustrated in Figs. 30I and J, for it may be tetrahedral, as in leptosporangiate ferns, or it may be cubical, as in the Eusporangiatae. The tapetum is formed from the outermost layers of the sporogenous tissue, unlike that of the Eusporangiatae, and there is also a layer of tabular cells, formed from the same regions, which becomes appressed to the inner side of the sporangium wall. For this reason, at maturity, the wall appears to be two cells thick. There is a primitive kind of annulus, formed by a group of thick-walled cells, on one side of the sporangium and a thin-walled stomium, along which dehiscence occurs, extends from it over the apex of the sporangium (Fig. 30F). Relatively large numbers of spores are released from each sporangium (e.g. about 128 in *Leptopteris* and more than 256 in *Osmunda* and *Todea*). The spores contain chlorophyll and must germinate rapidly if they are to do so at all.

The prothallus (Fig. 30C) is large, fleshy and dark green, resembling a thalloid liverwort, up to 4 cm long. The antheridia (Fig. 30D) project from the surface, as in Leptosporangiatae, but are larger, have more wall cells and produce a greater number of antherozoids than do most of them. The archegonia (Fig. 30E) are borne along the sides of the midrib; they have projecting necks and differ from those of leptosporangiate ferns only in the number of neck cells (six tiers, instead of the usual four).

The embryology of the young sporophyte, too, shows some

features which distinguish the Osmundales from the Leptosporangiatae. Not only is the first division of the zygote vertical, but so also is the second. It is the third division which is at right angles to the axis of the archegonium, instead of the second. Subsequent divisions are somewhat irregular and the embryo remains spherical for a relatively long time. Ultimately, however, a shoot apex, cotyledon, root and a large foot appear, but there is some irregularity in their derivation from the initial octants.

Despite the marginal position of the sporangia in *Osmunda*, as compared with their superficial position in the other two genera, the three genera are so similar in other respects that they are, without doubt, closely related, and this conclusion is supported by chromosome counts. The haploid number is $n = 22$ throughout.

FILICALES

In the past, the name Filicales was applied in the broadest possible sense, so as to include all the ferns but, recently, its use has been restricted, and it is applied just to the homosporous leptosporangiate ferns, as in Engler's *Syllabus der Pflanzenfamilien* (Melchior and Werdermann, 1954). However, even when thus restricted, it is still by far the largest group of the pteridophytes, for it contains almost 300 genera and about 9000 species. Details of their form and anatomy would occupy many volumes and can only briefly be summarized here, the following families, subfamilies and genera having been selected to illustrate the salient points.

Schizaeaceae *Senftenbergia*, Klukia*, Schizaea, Lygodium, Mohria, Anemia, Actinostachys*
Gleicheniaceae *Oligocarpia*, Gleichenites*, Gleichenia, Stromatopteris*(?)
Hymenophyllaceae *Hymenophyllum, Trichomanes, Polyphlebium*
Matoniaceae *Matonidium*, Matonia, Phanerosorus*
Dipteridaceae *Clathropteris*, Dictyophyllum*, Camptopteris*, Dipteris*

Cyatheaceae
 Dicksonioideae *Coniopteris*, Dicksonia, Cibotium*
 Cyatheoideae *Alsophilites*, Hemitelia, Cyathea, Alsophila*

Dennstaedtiaceae
 Dennstaedtioideae *Dennstaedtia, Microlepia*
 Pteridioideae *Pteridium, Histiopteris*
 Davallioideae *Davallia*
 Oleandroideae *Nephrolepis*

Onocleoideae(?)   *Onoclea, Matteuccia*
Blechnoideae   *Blechnum, Woodwardia*
Asplenioideae   *Asplenium, Phyllitis*
Athyrioideae   *Athyrium*
Dryopteridoideae   *Dryopteris, Polystichum*
Lomariopsidoideae   *Elaphoglossum*

Thelypteridaceae   *Thelypteris*
Adiantaceae   *Adiantum, Cheilanthes, Pellaea, Ceratopteris, Anogramma,*
         Acrostichum(?), *Pteris*(?)

Polypodiaceae   *Platycerium, Polypodium, Stenochlaena*(?)

As might be expected in such a large group, there is a considerable range of form and growth habit, from tiny annuals to tall tree-ferns and from protostelic forms to those with highly dissected polycyclic dictyosteles, yet all are alike in the early stages of development of the sporangium. This, together with its stalk, arises from a single cell. The first division of the initial cell (Fig. 31S) is into an apical cell (1) and a basal cell (2). Further divisions take place in each (Fig. 31T) and give rise to a primary sporogenous cell (shaded in Fig. 31U) and a jacket cell (3). The former gives rise to a two-layered tapetum and to a number of spore mother cells, surrounded by a sporangium wall one cell thick (Fig. 31W). Further details of sporangium development differ according to species, for some have a long slender stalk, only one cell thick, while others have a short and relatively thick stalk; the majority have a vertical row of thick-walled cells, constituting the annulus, while some have an oblique row and others merely a group of thick-walled cells; some have a high spore output, while in most species it is thirty-two or sixty-four.

Most commonly the prothallus is either cordate or butterfly-shaped ranging in size from a few mm to 1 cm or more across. There is a midrib several cells thick, but the wings of the prothallus are only one cell thick. It is surface-living, green and photosynthetic, and there are rhizoids on the underside, among which antheridia and archegonia are borne; the archegonia are usually concentrated near the growing point, or 'apical notch'. Departures from this typical form occur in certain families, e.g. some have filamentous prothalli, resembling an algal filament, while even subterranean prothalli are known, but this habit is extremely rare.

Stages in the development of the archegonium are illustrated in Figs. 31O–R, the only variations being in the number of tiers of neck cells at maturity. The precise way in which the antheridia of the Filicales develop has been the subject of some debate. Until 1951, the classical view had been widely accepted, according to which one or two funnel-shaped cell walls were said to be laid down. However, no attempt had been made by early workers to explain how this was possible. Having examined the process in some ten genera, Davie (1951) challenged this classical idea and put forward an alternative interpretation, as illustrated in Figs. 31B–G. According to him, successive cross-walls bulge upwards or downwards, so as to produce the characteristic ring-shaped cells of which the mature antheridium wall is constructed (Fig. 31H). However, Stone (1962) shows that the classical interpretation is correct for the three genera that she has examined. Thus, in *Polyphlebium*, whose antheridium has two ring

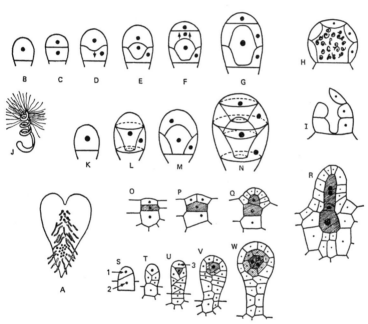

Fig. 31   Development of gametangia and sporangia in leptosporangiate ferns

A, typical gametophyte; B–H, stages in development of antheridium, according to Davie (diagrammatic); K–N, stages in development of antheridium, according to Stone; I, dehiscing antheridium; J. antherozoid of *Pteridium*; O–R, stages in development of archegonium; S–W, stages in development of sporangium of *Polypodium*. (1, apical cell; 2, basal cell; 3, jacket cell.)

(A, O–W, after Foster and Gifford; B–H, based on Davie; J, after Sadebeck; K–N, based on Stone.)

shaped cells, two funnel-shaped walls are laid down, as illustrated in Figs. 31K–N. At maturity, the cap cell is pushed off (Fig. 31I) to release the antherozoids (usually thirty-two in number) (Fig. 31J.) Some families have a slightly more massive antheridium, composed of a greater number of wall cells and containing more antherozoids; these are believed to be more primitive than the rest.

The embryology of the leptosporangiate ferns is likewise very constant throughout. The first cross-wall is almost invariably longitudinal and the second transverse. Thus, the zygote is divided at a very early stage into four quadrants, two directed towards the apical notch of the gametophyte (called the inner and outer anterior quadrants) and two away from the notch (called the inner and outer posterior quadrants). The outer anterior quadrant ultimately gives rise to the first leaf, the inner anterior to the shoot apex, the outer posterior to the first root, and the inner posterior to the foot. This, at least, is the procedure described in classical studies, but more recently it has been stated that the fate of the four quadrants is not always so clearly defined (Ward, 1954).

Statements that certain characters are primitive and others advanced, can be made with more certainty for the Filicales than for any other group in the plant kingdom, because of the large number of fossil representatives that are known. Some of the families had already become widespread by the Mesozoic, while others appeared as long ago as the Carboniferous. A comparison of these with the rest of the living Filicales makes it possible to draw up an extensive list of primitive characters for the group as a whole. The following list is based on that of Bower (1923), as modified by Holttum (1949), with additions by Stokey (1951).

Rhizome – slender, creeping, dichotomous, with fronds in two ranks on its upper side, protostelic, covered with hairs.

Fronds – large, amply branched, dichotomous and of unlimited growth, the stipe (petiole) receiving a single leaf trace, the ultimate pinnules narrow and with a single vein; venation without anatomoses (i.e. 'open').

Sori – containing few sporangia, terminating a vein.

Sporangia – relatively large, with stout stalk, without a specialized annulus, developing and dehiscing simultaneously to liberate a large number of spores.

Spore germination – giving a plate rather than a filament of cells.

Gametophyte – relatively large, thalloid, with a thick midrib, slow to develop.

Antheridium – large, containing several hundred antherozoids; wall cells more than four in number.

Archegonium – with a relatively long neck.

In the more advanced ferns, the dermal appendages are usually scales instead of hairs and, as the stem assumes an erect position, the leaves tend to form a crown at the apex. With increasing size, the stelar anatomy becomes more complex, the leaf-gaps overlap, and a dictyostele results. True vessels are known to occur in at least two genera (Eames, 1936). The fronds become reduced in size and may have a simple broad lamina with an entire margin and with anastomosing veins, while the stipe receives a number of leaf traces. In the most advanced ferns, the fronds are frequently 'jointed' at the base, i.e. they are shed by means of an absciss layer, a habit which may well be associated with life outside the tropics, in regions where seasonal changes in climate may be severe. Evolution of the sorus appears to have taken place in stages, the first of which involved a regular gradate sequence of development of the sporangia. The next resulted in a mixed arrangement of old and young sporangia within the sorus. Still more highly advanced is the condition described as 'acrostichoid', where the individuality of the sorus is lost and the sporangia form a 'felt' that covers the dorsal surface of the lamina, irrespective of the position of vein endings.

The various stages in soral evolution are often held to be the most important indicators of relative advancement and, on this basis, many pteridologists subdivide the Filicales into Simplices, Gradatae and Mixtae. It is important to realize, of course, that these subdivisions represent levels of evolution and *not* taxonomic groups. However, it is debatable whether one character should be weighted to this extent, for it is almost universally agreed among taxonomists that the maximum possible number of characters should be used in the assessment of phylogenetic status. If all the primitive characters listed above are taken into account, it is possible to calculate roughly an average 'advancement index' for each family or subfamily, ranging from 0 per cent (the most primitive) to 100 per cent (the most advanced) (Sporne, 1949). This has been done for the families and subfamilies selected for detailed treatment, and they have been arranged (Fig. 32) on a circular scheme, according to their advancement index. The most primitive families are near the centre and the most advanced are near the outside. The broken lines, enclosing 'areas of affinity', indicate which groups are most closely related to each other. During the past half century, views have fluctuated concerning phylogenetic relationships within the Filicales, and they are still in a state of flux. Thus, while Fig. 32 incorporates some of the opinions expressed recently in the symposium volume edited by Jermy, Crabbe and Thomas (1973), it must be expected that, as

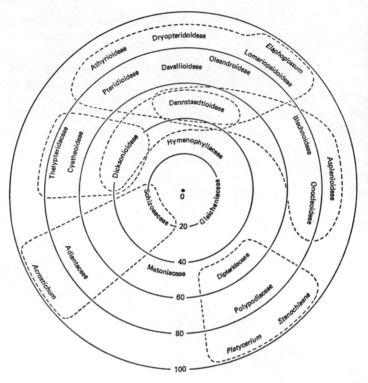

Fig. 32   Circular phylogenetic classification of the Filicales

Families and subfamilies are arranged so that their radial position corresponds to their relative advancement (primitive near the centre; advanced near the outside). Broken lines enclose 'areas of affinity', indicating close relationship. The numbers represent successive grades of relative advancement, expressed as a percentage ('the advancement index'), ranging from the most primitive (0%) to the most advanced (100%)

further facts are discovered, alterations will have to be made to it. Such a scheme may be thought of as a view, looking down from above, of the 'tree of evolution' of the Filicales, and while it may not be acceptable to all taxonomists, it does avoid the error, which is common to most phylogenetic classifications, of suggesting that one modern family has evolved from another modern family (Sporne, 1956).

The two most primitive familes are the Schizaeaceae and the Gleicheniaceae, and they are also the oldest, being represented in Carboniferous deposits by *Senftenbergia* and *Oligocarpia* respectively. Both are represented in the Mesozoic, too (viz. *Klukia* and *Gleichenites*).

SCHIZAEACEAE

The Schizaeaceae are represented today by five genera and about 160 species, most of which are tropical or subtropical in distribution. In all of them, the sporangia are borne singly instead of in sori ('monosporangial sori') and they show the most primitive type of dehiscence mechanism known in the Filicales. In all, the annulus consists merely of a terminal group of thick-walled cells (Figs. 33A–D) and dehis-

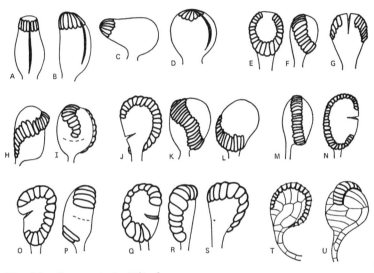

Fig. 33   Sporangia in Filicales

A, *Anemia*; B, *Schizaea*; C, *Lygodium*; D, *Mohria*; E–G, *Gleichenia*; H, I, *Matonia*; J–L, *Hymenophyllum*; M, N, *Cibotium*; O, P, *Hemitelia*; Q–S, *Dipteris*; T, U, *Asplenium*.

(A–D after Prantl; E–S, Bower; T, U, Müller.)

cence is longitudinal. The stalk of the sporangium is short and thick and the spore output from each is 128 or 256. The sporangia arise simultaneously, on the margin of the frond, and are unprotected, except by the inrolling of the margin, or marginal flaps, of the pinnule. *Lygodium* is one of the few modern genera of ferns to have fronds of unlimited growth, forming twining structures 30 m or more in length. Unlimited growth is a feature which, in most plants, is taken to distinguish stems from leaves. When it occurs in fronds, as in this case, it is, therefore, taken as evidence that they have evolved from stem structures (or are still in the process of doing so). Further evidence that the frond of *Lygodium* is very primitive is provided by the structure of the leaf trace, which shows only slight departures from radial symmetry. The other genera have leaf traces which are clearly dorsiventral and 'gutter-shaped'. Their stem structures, too, are more advanced for, whereas *Lygodium* has a creeping protostelic rhizome, *Schizaea* has an oblique rhizome with a medullated protostele, *Anemia* has a creeping or oblique rhizome which is either solenostelic or dictyostelic, while *Mohria* is dictyostelic. It is interesting to note, also, that *Mohria* is the most advanced in its dermal appendages, for they are glandular scales, whereas those of the other genera are hairs. In *Anemia*, only the two lowermost pinnae are fertile.

The prothalli are flat thalloid structures, except in *Schizaea*, in most species of which they are uniseriate filaments, and in *Actinostachys*, where they are colourless, subterranean and cylindrical, becoming tuberous with age (Bierhorst, 1966, 1968a). Those of *S. dichotoma* are subterranean and consist of branching multiseriate filaments (Bierhorst, 1967). Bower (1926) remarked that the filamentous prothalli of *Schizaea* are 'the simplest prothalli known among the Pteridophyta. They suggest a primitive state and provide comparison with green Algae.' However, their simplicity subsequently came to be regarded as the result of evolutionary specialization, instead of representing a primitive state. Bierhorst (1967) agrees that this may well be so in more advanced ferns, such as *Woodsia*, but regards the prothalli of *Schizaea* and *Actinostachys* as primitive. Moreover, he draws attention to similarities between the latter genus and the Psilotaceae. Not only are the prothalli similar, but so also are the early stages of development of the sporophyte, which are quite un-fernlike. Thus, the zygote of *Actinostachys* divides transversely, the inner cell develops into a foot and the outer one into a stem apex.

Of the five genera, *Lygodium* has the most complex antheridial wall and the highest output of antherozoids (156).

GLEICHENIACEAE

This family is represented by about 130 species belonging, mostly, to the one genus, *Gleichenia* (some taxonomists prefer to split the genus into four). A number of rather different types of leaf morphology occur (Holttum, 1954), two of which are illustrated in Figs. 34C and D, but in all of them the growth of the main rachis is arrested, until a pair of primary laterals has formed. In some species, these are of limited growth (Fig. 34C), but in others they too may terminate in dormant buds, so producing a variety of patterns, some looking superficially like a series of regular dichotomies (although, in fact, they are pseudo-dichotomies, because of the dormant apical bud in each angle). In others, there is a zig-zag arrangement of branches (Fig. 34D). As in the Schizaeaceae, therefore, the fronds are of indefinite growth, and some attain a length of 7 m or more. They arise from a creeping dichotomous rhizome which in most species is protostelic. A few, however, achieve a solenostelic condition, e.g. *G. pectinata*, a relatively advanced condition which is associated with a larger number of sporangia in the sorus than is usual in the genus. Yet, in this species, the dermal appendages are hairs, whereas scales are commonly present in others. Divergent facts such as these serve to emphasize the point that the evolution of different characters does not necessarily keep step, the result being that most organisms show a combination of advanced and primitive characters. This is why it is unwise to focus attention unduly on one character, when attempting to assess the relative advancement of taxonomic groups.

The sporangia, in strong contrast to those of the Schizaeaceae, are borne superficially on the abaxial side of the frond. They develop simultaneously and are arranged in sori containing, often, only a single ring of sporangia, seated either at a vein ending or, more usually, over the middle of a vein. There is no indusium at all covering the sorus, whose only protection is a covering of hairs or scales. Each sporangium is pear-shaped (Figs. 33E–G), has a stout stalk, and dehisces by means of an apical slit. Dehiscence is brought about by the contraction of the thickened cells of the annulus, which runs obliquely round the sporangium wall. Large numbers of spores are liberated from each, ranging from 128 to more than 1000.

The gametophyte is primitive, in that it is large, massive and slow growing. When old, it becomes much fluted and develops an endophytic mycorrhizal association. The antheridia are larger than in any other leptosporangiate fern and resemble those of the Osmundales. Those of *G. laevigata* are as much as 100μ in diameter and contain several hundred antherozoids.

Stromatopteris is represented by the single species *S. moniliformis*, which is endemic to New Caledonia, where it grows in the same habitat as *Schizaea dichotoma* and four species of *Actinostachys*. Bierhorst (1968 a and b, 1969, 1971) has made an extensive study of the morphology of this interesting species, in the light of which he believes that it should be removed from the Gleicheniaceae and placed in a separate family, Stromatopteridaceae. Perhaps the most remarkable feature of the sporophyte is that the frond, instead of being a lateral appendage on the stem, is continuous with it, there being a gradual transition from the one to the other. There is, thus, the problem, already encountered in many Palaeozoic ferns, of deciding where stem ends and frond begins, and it is of great interest, therefore, that among living ferns it should occur in such a primitive group.

HYMENOPHYLLACEAE

This group is commonly referred to as 'the filmy ferns', because of their delicate fronds, the lamina of which is usually only one cell thick. There are some 300 species of *Hymenophyllum*, of which two occur in the British Isles, and 350 of *Trichomanes*, of which there is one in the British Isles. Because of their delicate nature, almost all of them are confined to moist habitats, and most of them are restricted to the tropics, where they commonly grow as epiphytes. The British species *H. tunbrigense* may be seen growing on rocks constantly wetted by the spray from waterfalls.

Most filmy ferns have a thin wiry creeping, protostelic rhizome, from which the fronds arise in two rows. In one species, the stele of the rhizome is reduced to a single tracheid, while in another there is said to be no xylem at all. Some species are completely without roots. The leaf trace is a single strand, which at the base of the stipe shows marked similarity to the stem stele but, higher up the stipe, broadens

---

Fig. 34   Leaf form and sori of Filicales

A–D, leaf form: A, *Phanerosorus sarmentosus*; B, *Matonia pectinata*; C, *Gleichenia longissima*; D, *G. linearis*, var. *alternans*. E–W, sori: E, *Matonia pectinata*; F, *Trichomanes alatum*; G, *Cibotium barometz*; H, *Cyathea Dregii*; I, *Dennstaedtia cicutaria*; J, *Microlepia speluncae*; K, *Matteuccia struthiopteris*; L, *Dryopteris filix-mas*; M, *Polystichum lobatum*; N, *Nephrolepis davallioides*; O, *Pteris tripartita*; P, *Pteris cretica*; Q, *Pteridium aquilinum*; R, *Athyrium filix-foemina*; S, *Adiantum Parishii*; T, *Lomaria spicant*; U, *Blechnum occidentale*; V. *Phyllitis scolopendrium*; W, *Asplenium lanceolatum*.
(1, outer indusium; 2, inner indusium.)

(A, B, J, T–V, after Diels; C, D, O, Holttum; F, Bauer; G, H, S, Hooker; I, Baker; K, L, M, W, Luerssen; N, R, Mettenius; P, Q, Bower.)

Fig. 34

out into a gutter-shaped strand. The frond is usually much branched, each narrow segment having a single vein, but various degrees of 'webbing' occur and, in one species, *Cardiomanes reniforme* (= *Trichomanes reniforme*), there is a single expanded lamina. Nevertheless, the venation is open in all species.

The sori are marginal, and most species are strictly gradate. The vein leading to the sorus continues into a columnar receptacle which in *Trichomanes*, can grow by means of an intercalary basal meristem until it forms a slender bristle. The receptacle of *Hymenophyllum* has more limited powers of growth or may lack them altogether. In such species, the sporangia are produced simultaneously, but, where the receptacle can grow, new sporangia arise in basipetal sequence. Surrounding the sorus is a cup-shaped indusium in *Trichomanes* (Fig. 34F, where the broken line indicates where the indusium was cut away to show the base of the receptacle) and a two-lipped indusium in *Hymenophyllum*.

The sporangium has a relatively thin stalk and an oblique annulus, which brings about dehiscence along a lateral line (Figs. 33J–L), by a process of slow opening, followed by rapid closure as a gas phase suddenly appears in the cells of the annulus. This mechanism is found throughout the more highly evolved members of the Filicales, and results in the forcible ejection of the spores. The spore output varies from 128 or 256 in *Hymenophyllum* to as low as thirty-two in some species of *Trichomanes*.

The prothallus of *Hymenophyllum* is a strap-shaped thallus, often only one cell thick, but, by contrast, many species of *Trichomanes* have a filamentous structure which, like that of *Schizaea*, is mycorrhizal.

Recent work (Farrar and Wagner, 1968) suggests that, in some species of *Trichomanes*, the form of the gametophyte may be influenced by environmental conditions. Thus, under reduced light the prothallus of *T. holopterum*, which is normally plate-like, produces filamentous outgrowths.

MATONIACEAE

This is a most interesting family, containing the two genera *Phanerosorus*, from Sarawak and New Guinea, and *Matonia*, from Malaya, Borneo and New Guinea. In spite of its rarity at the present day, the family had many fossil representatives in the Triassic. So characteristic is the method of branching of the frond (Fig. 34B) that there can be little doubt that the fossil *Matonidium* is correctly placed in this family. After an initial dichotomy, each half of the frond undergoes a regular series of unequal catadromic dichotomies (i.e. each takes the main growing point further from the median plane). Each pinna is pinnatifid and there are anastomoses in the veinlets, particularly in the neighbourhood of the sori. *Phanerosorus* (Fig. 34A) has a frond of indefinite growth which is long and slender and bears dormant buds at the tips of some of its branches.

The stem of *Matonia* is creeping and hairy, and has a very characteristic polycyclic stelar structure, with two co-axial cylinders surrounding a central solid stele. From these, a single gutter-shaped leaf-trace is formed, both cylinders playing a part in its origin.

The sori are superficial and consist of a small number of sporangia arranged in a ring round the receptacle, which continues into the stalk of an umbrella-shaped indusium (Fig. 34E represents a vertical section through a young sorus). There is an oblique convoluted annulus round the sporangium, dehiscence being lateral, although there is no special stomium of thin-walled cells (Figs. 33H and I). The spore output is sixty-four.

DIPTERIDACEAE

This family is represented at the present day by some eight species of the single genus *Dipteris*, restricted to the Indo-Malayan region, but in Triassic times there were at least three genera, *Clathropteris*, *Dictyophyllum* and *Camptopteris*. Again, the architecture of the leaf is quite characteristic, and there can be little doubt as to the correct taxonomic placing of these fossil forms. After an initial dichotomy, the frond shows successive unequal dichotomies in an anadromic direction (i.e. towards the median plane). This pattern is represented in present-day species, in the venation of the two halves of the frond. However, while the primary veins are dichotomous, the smaller ones form a reticulum of a highly advanced type, with blind-ending veinlets, as in the leaves of many flowering plants.

The fronds arise at distant intervals along a creeping hairy rhizome, whose vascular structure is a simple solenostele. While some species have only a single leaf trace, others have two entering the base of the stipe.

The sorus is superficial, completely without an indusium, and the sporangia are interspersed with glandular hairs. In *Dipteris Lobbiana* the sporangia arise simultaneously, but in *D. conjugata* they are mixed. Thus, the single genus cuts right across the division of the ferns into Simplices, Gradatae and Mixtae.

The sporangia have relatively thin stalks (only four cells thick) the annulus is oblique (Fig. 33Q–S), and dehiscence is lateral. The spore output is sixty-four.

Most large tree ferns belong either to the genus *Cyathea* (*sensu lato*) or to *Dicksonia*. In the early days of fern taxonomy, they were united into the family Cyatheaceae. Bower (1923), however, segregated them into two distinct families, Cyatheaceae and Dicksoniaceae, stressing the importance of the superficial position of the sori in the former and their marginal position in the latter. He believed that the two families have had quite separate evolutionary histories from quite different ancestors with superficial and marginal sori respectively. Certainly, they were distinct from one another as far back as the Jurassic. Holttum (1949) was in favour of this separation, but he has now changed his mind (Holttum and Sen, 1961) and he believes that the two families should be merged into the Cyatheaceae once more. It is, nevertheless, convenient to retain the distinction between the two groups at the level of the subfamily between the Dicksonioideae and the Cyatheoideae.

The first recorded occurrence of a fossil member of the Dicksonioideae is of *Coniopteris*, from Jurassic rocks of Yorkshire. Like modern members of the group, it had highly compound fronds with marginal sori, protected by two flaps (the upper and lower indusia). In the modern genus *Cibotium*, the fronds are borne on stout creeping stems or on low massive trunks, while some species of *Dicksonia* are tall tree-ferns (e.g. *D. antarctica*), with a crown of leaves at the summit of a tall trunk. All are characterized by a profuse hairy covering over the stem and the base of the stipe, the hairs being as much as 2 cm long in *Cibotium barometz*.

The stems are solenostelic or (in species with erect axes) dictyostelic, and the stele is deeply convoluted around a large central pith region. There is a single gutter-shaped strand entering the base of the stipe, but this soon breaks up into numerous small bundles.

The sporangia are truly marginal in origin and arise in strictly gradate sequence within a purse-like box, formed by the two indusia (Fig. 34G). They are long-stalked and have an oblique annulus

(Figs. 33M and N) which, in some species, is very nearly vertical. The typical spore output per sporangium is sixty-four.

CYATHEOIDEAE

The earliest known fossil representative of the group is *Alsophilites* from the Jurassic. Bower (1926) recognized three living genera within the family: *Alsophila* with about 300 species, *Hemitelia* with about 100, and *Cyathea* with about 300. Holttum and Sen (1961), however, regard the distinction between these three genera as artificial, and prefer to merge them into the one genus *Cyathea*.

Although the largest may attain a height of 24 m, some species are comparatively low-growing. Much of the diameter of the trunk is composed of matted adventitious roots and persistent leaf bases, while the stem within is relatively small. Nevertheless, its stelar anatomy is highly complex for, in addition to a convoluted dictyostele, there are abundant medullary strands, and sometimes cortical strands too. Broad chaffy scales form a dense covering over the stem apex and the base of the frond.

The stipe receives a number of separate leaf-traces from the lower margin of the associated leaf gap. While the fronds of most species are several times pinnate, those of *Cyathea sinuata* are simple. The venation is open in the majority of species, except for very occasional vein fusions.

The three genera recognized by Bower are distinguished by the character of the indusium but, otherwise, the sori are very similar in their gradate development. In *Alsophila* there is no indusium at all, in *Hemitelia* there is a large scale at one side of the receptacle, and in *Cyathea* (Fig. 34H) it extends all round the receptacle to form a cup which completely covers the globose sorus when young, but which becomes torn as the sporangia develop and push through it.

Bower suggested that *Cyathea* has affinities with *Gleichenia* and that its cup-like indusium might have evolved from the hairs which surround the receptacle of some species of *Gleichenia*. The indusium of *Hemitelia* then represents a stage in the reduction process that has led to the complete absence of an indusium in *Alsophila*. Holttum and Sen (1961), however, visualize the *Cyathea*-type of indusium as having evolved *from* the *Hemitelia*-type by extension of the scale-like indusium so as to surround the sorus completely. The *Hemitelia*-type of indusium, in turn, they compare to the inner indusium of *Dicksonia*, suggesting that the position of the sorus has become progressively more superficial during the evolution of the Cyatheoideae from the Dicksonioideae.

The sporangium is relatively small, with a four-rowed stalk, an oblique annulus (Figs 33O and P), and a fairly well marked lateral stomium. The spore output ranges from sixty-four to sixteen, and even eight in some species.

DENNSTAEDTIACEAE

We now come to the large assemblage of ferns whose sori show the mixed condition and which Bower grouped together in the one big artificial family, the Polypodiaceae. Some, he believed, had affinities with the Dicksoniaceae, some with the Cyatheaceae and some with the Osmundaceae, yet all had achieved the same advanced type of sporangial structure, with a thin stalk, a vertical incomplete annulus, and lateral dehiscence. Figs. 33T and 33U are two views of the sporangium of *Asplenium*, which demonstrate the small number of cells constituting the capsule, and the way in which the stalk is composed of just one row of cells, in the most highly evolved types.

In 1949 Holttum suggested a more nearly natural classification of these ferns, by creating a new family, the Dennstaedtiaceae, within which he grouped a number of subfamilies which, he believes, have affinities with the Dicksoniaceae. In this new scheme of classification, the Polypodiaceae constitute a very restricted family, having affinities with the Matoniaceae, the Dipteridaceae being absorbed into it. Within the Dennstaedtiaceae, so many evolutionary processes have taken place that the group is hard to define; indeed, it would almost seem that the subfamilies warrant elevation to family status.

DENNSTAEDTIOIDEAE

This is the most primitive of the subfamilies of the Dennstaedtiaceae, for some species still retain the gradate arrangement of sporangia in the sorus. Most have creeping rhizomes with solenosteles. The sorus of *Dennstaedtia* (Fig. 34I) is very similar indeed to that of *Dicksonia* in having two indusia. In *Microlepia*, however, the upper indusium is greatly expanded (Fig. 34J), so that, in spite of its marginal origin, the sorus appears to be superficial at maturity. This represents an early stage in the evolutionary process which Bower called the 'Phyletic Slide', whereby the sorus ultimately has a superficial origin despite its marginal ancestry.

DAVALLIOIDEAE

Davallia, likewise, has a superficial sorus at maturity, covered by a funnel-shaped indusium, but which, nevertheless, is marginal in origin. The stem is creeping, with a peculiar type of dissected solenostele, and is clothed with scales.

OLEANDROIDEAE

Nephrolepis has upright, dictyostelic stems with long runners, by means of which vegetative reproduction occurs, for the tips of the runners are capable of rooting and turning into normal erect stems. Within the genus, there is a wide range of soral form. *N. davallioides* (Fig. 34N) is very similar to *Microlepia*, in that the upper indusium is scarcely larger than the lower. In *N. acuta*, the sorus is superficial, not only at maturity, but also in origin. *N. dicksonioides* shows a different evolutionary trend, in that adjacent sori are sometimes fused, and this trend has proceeded so far in *N. acutifolia* that the margin of the pinna has a sorus running continuously along it, between two linear indusia.

PTERIDIOIDEAE

It is generally accepted that the sorus of *Pteridium* evolved in a similar way to that of *Nephrolepis acutifolia*, for it, too, is continuous along the margin of the pinnule (Fig. 34Q) and is protected by two indusia. The upper indusium (1) is relatively thick, but the lower one (2) is thin and papery. *Pteridium* is one of the most successful ferns in its ability to compete with flowering plants and this may, to some extent, be due to the great depth at which its rhizomes spread beneath the surface of the soil. Its stele is a dicyclic perforated solenostele. *Histiopteris* has a much simpler, unperforated solenostele (Campbell, 1936), but its soral arrangement is regarded as more advanced than that of *Pteridium* for it lacks an inner indusium.

ONOCLEOIDEAE

Holttum leaves this subfamily unplaced in his classification, while Bower thought that it shows some affinities with the Cyatheaceae and with the Blechnoideae. It contains two genera, *Matteuccia* (two species) and *Onoclea* (monotypic). Both are markedly dimorphic, with specially modified fertile fronds. The fertile pinnae are narrow and the margins are tightly inrolled so that protection of the sorus is derived more from them than from the indusium, which is thin and papery (Fig. 34K). Both are dictyostelic and covered with scales. *Matteuccia* has open venation and *Onoclea* reticulate.

DRYOPTERIDOIDEAE

The ferns in this subfamily have a short stout stem which is more or less erect, dictyostelic and covered with scales. The stipe receives numerous leaf traces, and the venation is open. The sori are superficial on the veins, or at vein endings, and are covered by an indusium which in *Dryopteris* is reniform (Fig. 34L) and in *Polystichum* is peltate (Fig. 34M). Of these the reniform type is probably the more primitive, for it is not far removed from the condition figured for *Nephrolepis* (Fig. 34N). From this type, it is easy to imagine the evolution of the radially symmetrical indusium of *Polystichum*, by the extension of the 'shoulders' round the point of attachment, followed by a 'fusion' to form a disc, with a central point of attachment.

ATHYRIOIDEAE

Some species of *Athyrium* have indusia that are identical in shape with those of *Dryopteris*, but most have two types on the same frond, as does the British *A. filix-femina* (Fig. 34R). Here, there are some sori with reniform indusia and some in which the indusium is extended along the lateral veins. The vascular supply to the stipe of the frond consists of two leaf traces, which unite into a single gutter-shaped strand higher up.

LOMARIOPSIDOIDEAE

All the members of this subfamily are acrostichoid. There has been much discussion as to their affinities, but they probably lie with the Davallioideae, for the stele of *Elaphoglossum* is very similar to that of *Davallia*, in having two large meristeles connected into a cylinder by a network of smaller bundles.

ASPLENIOIDEAE

This subfamily, too, is believed by Holttum to have affinities with the Davallioideae. The sorus of *Asplenium* (Fig. 34W) is extended along the lateral veins and is protected by an indusium which is usually acroscopic (i.e. its free margin is directed towards the apex of the pinna). In this, it resembles most of the sori of *Athyrium*. However, the vascular supply to the stipe is different from that in *Athyrium*, for the two bundles which enter it fuse into a single four-armed strand, instead of into a gutter-shaped strand. The same is true of *Phyllitis*. That *Asplenium* and *Phyllitis* are closely related seems fairly certain, since they have the same basic chromosome number, $n = 36$, and hybrids between them are known to occur. In *Phyllitis* the sori occur in pairs, facing each other, along the lateral veins (Fig. 34V), one acroscopic and the other basiscopic.

BLECHNOIDEAE

Blechnum punctulatum forms a possible intermediate between *Phyllitis* and the more typical species of *Blechnum*, for on one and the same frond both types of sorus may occur, some in pairs facing each other and some showing various degrees of fusion along a commissural vein. *Woodwardia* has a series of box-like sori, on either side of the midrib, whose indusia are like hinged lids. The typical *Blechnum* sorus is a continuous one, as if the adjacent sori of a *Woodwardia* had become fused together, with the indusium facing the midrib of the pinna (Fig. 34U). Each has beneath it a commissural vein, which is visible in Fig 34T, where part of the two sori have been removed to expose it. The British species, here figured, shows a considerable reduction of the fertile lamina, and this reduction process has gone much further in other members of the subgenus *Lomaria*, where the lamina is almost completely lacking. Such species are markedly dimorphic, for the sterile fronds have a normal unreduced lamina. The genus shows a wide range of habit, for some species are creeping, some are climbing, while several have erect trunks, like small tree ferns.

THELYPTERIDACEAE

A large number of ferns which have for a long time been confused with members of the Dryopteroideae have recently been segregated by Holttum (1971) and placed in a family of their own, which he calls the Thelypteridaceae. He believes them to be related to the Cyatheaceae. In the British Isles, there are three species of *Thelypteris*, among which is the very interesting Marsh Fern, *T. palustris*. It has a creeping rhizome which may have either a perforated solenostele or a perforated dictyostele. The two types of stele grade into one another according to the distance between successive fronds. This species thus provides a clear demonstration of the fact that the difference between solenosteles and dictyosteles is merely one of degree. In this species, the sorus is protected by a reniform indusium which is thin and papery. Indeed, it is more like that of *Hemitelia* than that of *Dryopteris*. Of the other two British species, one sometimes lacks an indusium and the other always lacks one.

ADIANTACEAE

This is a very diverse family, some members of which show marked similarities with *Mohria* (Schizaeaceae). Their sori are without indusia and occur along the veins or else form 'fusion sori' near the margin. *Adiantum* has the sporangia restricted to the underside of special reflexed marginal flaps of the lamina (Fig. 34S). The majority of the members of the family inhabit fairly dry regions and some are markedly xeromorphic, e.g. *Cheilanthes* and *Pellaea*. However, at the other extreme, *Ceratopteris* is a floating, or rooted, aquatic plant, now widespread in tropical countries, where it chokes up canals and slow moving rivers. *Anogramma leptophylla* is interesting, in having a subterranean perennial prothallus, from which arise delicate annual sporophytes.

In both *Cheilanthes* and *Pellaea*, the sporangial regions are protected by the inrolled margin of the lamina (sometimes called an 'indusium'). In *Pellaea* it is continuous, but in *Cheilanthes* it is interrupted. *Pteris* also has a continuous sorus which is protected by the inrolled margin of the lamina, as shown in Fig. 34O. In some species, e.g. *Pteris cretica*, the soral region is somewhat expanded, as shown in Fig. 34P, and indicates a possible way in which the acrostichoid condition, as exhibited by *Acrostichum* might have evolved. At one time, it was thought that *Pteris* might have evolved from *Pteridium* by the disappearance of the inner indusium, as suggested for *Histiopteris*. However chromosome counts do not support this view. Instead, they give strong support to the view that both *Pteris* and *Acrostichum* have affinities with the Adiantaceae whose basic chromosome numbers are 29 and 30; the basic number for *Pteris* is 29, and that for *Acrostichum* is 30 (Chiarugi, 1960; Fabbri, 1963, 1965).

Platyzoma, a monotypic genus confined to Australia, was at one time placed in the Gleicheniaceae, but is now believed to have affinities with *Ceratopteris*. Perhaps, however, it should be placed in a family of its own, as suggested by Tryon (1964) in view of its heterospory, which is unique among the Filicales. It produces megaspores which are about twice the size of its microspores. The small spores give rise to filamentous prothalli, bearing only antheridia; the larger spores produce spathulate prothalli with archegonia when young, but with antheridia later on, if fertilization has not occurred. This behaviour is closely comparable to that of *Equisetum*.

POLYPODIACEAE

Within this family are placed a number of genera of ferns, all of which completely lack any kind of indusium. There are about 1000 species in the family, almost all tropical in distribution (but note that *Polypodium vulgare* occurs in the British flora), and most are epiphytic. Many have highly complex anastomosing venation and some are acrostichoid, e.g. *Platycerium*. This genus is markedly dimorphic, with 'nest leaves' appressed to the tree trunk on which it is growing, while the fertile fronds are quite different in shape and give rise to the name 'Stag's horn fern'.

Stenochlaena is an acrostichoid genus which is believed by some to be closely related to *Acrostichum*, however, its basic chromosome number of 37 suggests that its affinities are more likely to be with the Polypodiaceae, where the commonest numbers are 36 and 37.

It is generally agreed that the Polypodiaceae are closely related to the Dipteridaceae and that their sori have been without an indusium throughout their evolutionary history.

It will be clear, from this brief survey of the Filicales, that there is much scope for disagreement among pteridologists as to the relationships and detailed phylogeny of the group, and that much more research is necessary before final conclusions can be reached. On the evidence so far available, it would seem that the group might well be diphyletic, with two evolutionary starting points, one with marginal and the other with superficial sori. Furthermore, it seems clear that even among those with marginal origins, there has been a trend towards the superficial condition. Should the Filicales prove to have been monophyletic, however, then it is most probable that the ancestral type had marginal rather than superficial sori, and that the Superficiales underwent a 'phyletic slide' early in their evolution, while the Marginales are proceeding more slowly in the same direction.

'WATER FERNS'

There are two interesting groups of leptosporangiate ferns which, at one time, were classified together as the Hydropterideae. Features which they show in common are heterospory and a hydrophilous habit, but in other respects they are so different as to warrant a much wider separation, from each other and from the rest of the ferns. Accordingly, their taxonomic status has been elevated to the Marsileales and the Salviniales respectively.

MARSILEALES

Pilulariaceae　*Pilularia*
Marsileaceae　*Marsilea, Regnellidium*

All the members of the Marsileales have creeping rhizomes, bearing erect leaves at intervals, on alternate sides. The only member of the group represented in the British flora is *Pilularia globulifera* ('Pill-wort'). Like all species of *Pilularia*, its leaves are completely without any lamina (Fig. 35A). The leaves of the monotypic Brazilian genus *Regnellidium* have two reniform leaflets. *Marsilea* occurs in temperate and tropical regions, many of its sixty-five species occurring in Australia. Its leaves have four leaflets and somewhat resemble a 'four-leaved clover' (Fig. 35B). All have solenostelic rhizomes but, in *Pilularia*, the vascular structure is much reduced, and the internal endodermis may be missing.

The sporangia, in all three genera, are borne in hard beanlike sporocarps, attached either to the petiole, near its base, or in its axil, either stalked or sessile. The morphological nature of these sporocarps has been the subject of much discussion, but it can most conveniently be regarded as a tightly folded pinna (like a clenched fist) enclosing a number of elongated sori, each covered by a membranous indusium. Figs. 35C and D represent, very diagrammatically, the structure of the sporocarp of *Marsilea* as seen in vertical and horizontal sections, respectively (for clarity, the number of sori has been reduced to two rows of five). Each receptacle bears microsporangia laterally and megasporangia terminally, and receives vascular bundles from a number of strands running down in the wall of the sporocarp. Arching over the top, is a gelatinous structure sometimes called a 'sporophore' (cross-hatched in the figures) which swells up at maturity and drags the paired sori from the sporocarp, as it dehisces (Fig. 35E).

The sporocarp of *Pilularia* is similar in construction, except that there are only four sori.

The sporangia are typically leptosporangiate in origin, the sporangium wall is very thin, and there is a tapetum of two or three layers

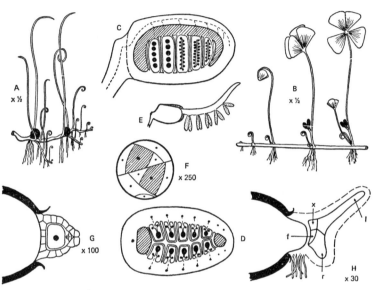

Fig. 35   Marsileaceae

A, *Pilularia globulifera*, growth habit. B–H, *Marsilea*: B, growth habit of *M. quadrifolia*; C, vertical section of sporocarp (simplified); D, horizontal section of sporocarp (simplified); E, dehiscing sporocarp; F, male gametophyte; G, female gametophyte of *M. vestita*; H, embryo within female gametophyte.
(f, foot; 1, leaf; r, root; x, stem apex.)

(B, after Meunier; C, D, based on Eames; E, after Eames; F, Sharp; G, Campbell; H, Sachs.)

of cells. The microsporangia contain thirty-two or sixty-four microspores but, in the megasporangia, all but one of the potential spores degenerate. On dehiscence of the sporocarp, the delicate sporangium wall rapidly decays and the spores begin to germinate almost at once.

The male gametophyte (Fig. 35F) is extremely simple, as in most heterosporous plants, consisting of nine cells only. There is a single small prothallial cell, and six wall cells surround two spermatogenous cells (cross-hatched in the figure) that give rise to sixteen antherozoids each.

The first cross-wall in the germinating megaspore is excentrically placed and cuts off a small apical cell, whose further divisions give rise to a single archegonium (Fig. 35G) with a short neck of two tiers of cells, one neck canal cell, a ventral canal cell and a large egg cell.

The first division of the zygote is longitudinal, and the second transverse, giving four quadrants (Fig. 35H) of which the outer two develop into the first leaf (1) and the first root (r), while the inner two develop into the stem apex (x) and the foot (f). Meanwhile, the venter of the archegonium grows and keeps place, for a time, with the enlarging embryo so as to form a sheath round it, from the underside of which a few rhizoids may be produced. The first few leaves in *Marsilea* are without a lamina and, therefore, closely resemble the leaves of *Pilularia*.

It is often claimed that this interesting group of ferns represents an evolutionary offshoot from an ancient schizaeaceous stock. Arguments for this are based on the leaf form, the type of hairs, the form of the sorus and the vestigial annulus round the apex of the sporangium in *Pilularia* (Eames, 1936), but the evidence is not very convincing and, in the absence of early fossil representatives, the group must be regarded in the meantime as an isolated one.

SALVINTALES

Salviniaceae   *Salvinia*
Azollaceae   *Azolla*

Whereas most of the members of the Marsileales are rooted in the soil, either in or near water, all the members of the Salviniales are actually floating. *Azolla* has pendulous roots, but *Salvinia* is completely without them. Like the Marsileales, their sporangia are borne in sporocarps. However, the morphological nature of the sporocarps is quite different, for each sporocarp represents a single sorus whose indusium forms the sporocarp wall.

The only member of the group to be represented in the British flora is *Azolla filiculoides*, described as recently naturalized from N. America. However, it was a native British plant in Interglacial times (West, 1953). It has an abundantly branching rhizome, with a minute medullated protostele, and with crowded overlapping leaves about 1 mm long (Fig. 36A). These have two lobes, within the upper of which is a cavity containing the blue-green alga *Anabaena azollae*.

Sporocarps arise on the first leaf of a lateral branch and are usually of two kinds – large ones containing many microsporangia and small ones containing a single megasporangium, although sporocarps with both types of sporangium are sometimes present. The early stages of development are similar in both types of sporocarp, for there is an

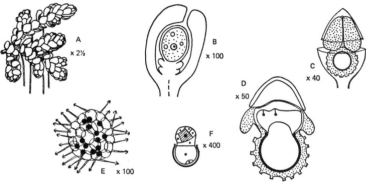

Fig. 36   Azolla

A, *Azolla filiculoides*, portion of plant; B, sporocarp with young megasporangium; C, megasporangium with massulae; D, the same, with female prothallus; E, massula from microsporangium; F, male prothallus.

(B, based on Pfeiffer; C, E, after Bernard; D, Campbell; F, Belajeff.)

elongated receptacle on which numerous sporangial initials arise. However, during development, the microsporangial initials abort in the one case (Fig. 36B) and the megasporangial initials abort in the other. In both types of sporangium there is an abundance of mucilaginous 'periplasmodium', which becomes organized into 'massulae'. In the megasporangium there are four such massulae, in one of which the single megaspore is buried. Fig. 36C illustrates the dehiscence of a megasporangium, the apex of which is cast adrift as a cap over the four massulae. The megaspore then germinates to produce a cap of prothallial tissue, within which several archegonia develop (Fig. 36D).

When the microsporangium dehisces a variable number of spherical frothy massulae are liberated, each with several microspores near the periphery. Each bears a large number of peculiar anchor-like 'glochidia' (Fig. 36E). These become entangled with the massulae surrounding a megaspore and, together, they sink to the bottom, where the microspores germinate, without being released from the massulae. The male gametophyte (Fig. 36F) has a single antheridium from which eight antherozoids are liberated.

The cleavage of the zygote is typical of the leptosporangiate ferns, and as soon as the first leaf appears the sporeling rises, carrying the massulae etc. once more to the surface.

There are about twelve species of *Salvinia*, several of which occur in Africa. Its horizontal floating stems, up to 10 cm long, have a much reduced vascular anatomy and bear leaves in whorls of three (Fig. 37A), two floating and one submerged. Whereas the floating leaves are entire and covered with peculiar unwettable hairs, the submerged leaves are finely divided into linear segments that bear a striking resemblance to roots (Fig. 37B). However, it is doubtful whether they perform the functions of roots. Growth is rapid and fragmentation occurs easily, with the result that ponds and lakes in tropical regions may rapidly become covered and canals choked.

The first few sporocarps to be formed in each cluster contain megasporangia, up to twenty-five in each, and the later ones microsporangia, in large numbers, on branched stalks (Fig. 37C). All except one of the potential megaspores in each megasporangium

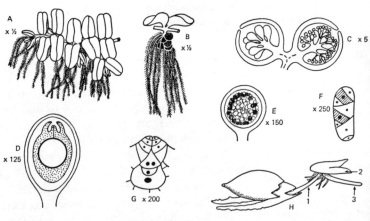

Fig. 37   Salvinia

A, *Salvinia natans*, portion of plant; B, node with sporocarps;
C, sporocarps with megasporangia and microsporangia; D, megasporangium; E, microsporangium; F, male prothallus; G, archegonium; H, female prothallus with sporeling attached.
(1, column; 2, leaf; 3, stem.)

(A, B, after Bischoff; C, Luerssen; D, H, Lasser; E, G, based on Yasui; F, after Belajeff.)

abort, and the functional megaspore becomes surrounded by a thick perispore (Fig. 37D), which later becomes cellular and comes to look, superficially, like the integument of a gymnosperm seed.

Within the microsporangium, the sixty-four microspores come to lie at the periphery of a single frothy massula (Fig. 37E). They remain within the sporangium throughout and, as they germinate, the male prothalli project all round. Each male prothallus contains two antheridia (Fig. 37F) producing a total of eight antherozoids.

The megaspore, too, remains throughout within the sporangium, after it has become detached. The female prothallus protrudes, as a cap of tissue from which extend backwards two narrow horizontal wings, or 'stabilizers'. Several archegonia develop, in a row, across the upper side of the projecting cap, each with a short neck, a neck canal cell with two nuclei, and a ventral canal nucleus (Fig. 37G).

Fig. 37H illustrates a young sporeling, still attached to the female prothallus within the megasporangium, and shows the peculiar development of a 'column' (1), separating the first leaf (2) and the stem (3) from the foot, which remains embedded in the prothallus. The early stages of segmentation of the zygote are not fully established. At no stage is a root primordium distinguishable.

If the relationship of the Marsileales are obscure, those of the Salviniales are even more so. The gradate origin of the sporangia within the sporocarp, the intercalary growth of the receptacle in *Azolla* and the vestigial oblique annulus have led to the suggestion that the group has affinities with the Hymenophyllaceae. However, this hardly seems acceptable, in view of the many extraordinary features that mark them off from all other ferns.

7 Progymnospermopsida

Extinct plants, mostly of Middle and Upper Devonian age, with certain anatomical characters normally associated with gymnosperms, but without seeds. Some known to have been tall trees with stout trunks made up of dense secondary wood. Ultimate branch systems either naked or bearing small lateral appendages showing varying degrees of flattening. Homosporous or heterosporous.

1 Aneurophytales*   *Aneurophyton** (+ *Eospermatopteris**?),
                           *Rellimia** (= *Protopteridium*, = *Milleria*),
                           *Tetraxylopteris** (= *Sphenoxylon*?),
                           *Triloboxylon**, *Proteokalon**

2 Protopityales*   *Protopitys**

3 Archaeopteridales*   *Archaeopteris** (+ *Callixylon**), *Pitys**(?)
                         (= *Archaeopitys*?)

This group was first proposed by Beck (1960) to accommodate a wide range of fossil remains of plants that shared certain anatomical features with gymnosperms, but which are believed, nevertheless, to have been at the level of pteridophytes in their reproductive processes. Some of them were originally thought to have had large fern-like fronds, and were indeed given generic names that were intended to suggest affinities with the ferns, e.g. *Protopteridium* and *Archaeopteris*. Subsequently, however, it has been shown that the 'fronds' were merely branch systems which, in life, were arranged in one plane like those of many modern conifers, or else became flattened during fossilization. The branch systems of others bore little resemblance to a modern fern frond and were three-dimensional. Names such as *Archaeopteris* are, therefore, misleading for the progymnosperms are not now believed to be at all closely related to ferns. Indeed, in their ability to become tall trees with dense secondary wood they were unlike all other pteridophytes. They differed from them also in having secondary phloem for, as Scheckler and Banks (1971a) emphasize, cambial activity in woody ferns, lycopods and calamites never produced any secondary phloem.

ANEUROPHYTALES

The first member of the progymnosperms to appear in the fossil record, from the lowermost Middle Devonian, was *Rellimia*. Originally known as *Protopteridium*, it was redefined under the name *Milleria* when the structure of its fertile regions was elucidated by Leclercq and Bonamo (1971) but, as this name was already occupied, it was subsequently changed yet again, to *Rellimia* (Leclercq and Bonamo, 1973). Fig. 38A is based on a reconstruction by Kräusel and Weyland (1933) and shows the spiral arrangement of sterile and fertile and branches. The latter were originally thought to show a slight development of a photosynthetic lamina. This was shown to be wrong, however, when Leclercq and Bonamo used the technique of *dégagement* for, on dissecting away the rock, they were able to show that each fertile appendage dichotomized at least once near the base (see Fig. 38B) and bore two orders of pinnate unflattened branches. Each ultimate branchlet terminated in a pair of sporangia. The entire system was curved adaxially and the sporangia (not shown in Fig. 38B) were on the concave side. The internal anatomy of *Rellimia* is not known, because of poor preservation, but there are indications of a lobed solid rod of primary xylem, surrounded by secondary wood made up of tracheids and wood-rays.

Aneurophyton, described by Kräusel and Weyland (1923, 1929), is widely regarded as synonymous with *Eospermatopteris*, described by Goldring (1924) from Upper Devonian deposits in New York state. The latter had a trunk estimated to have been at least 13 m tall and up to 60 cm in diameter. It was described as having had fronds up to 3 m long, pinnately constructed and bearing minute forked pinnules (Fig. 38C). *Aneurophyton* (Fig. 38D), was likewise described as a frond whose pinnae were all in one plane. However, its internal anatomy suggests otherwise, for the primary xylem consisted of a three-lobed solid rod of tracheids with a single mesarch protoxylem in each lobe. This is more consistent with a spiral arrangement of lateral branches which may subsequently have come to lie in one plane during the process of fossilization. The ultimate appendages, which forked once, twice or even three times, are now regarded as leaves, despite their reported lack of a vascular strand. The fertile appendages have not been subjected to detailed study by degaging techniques and are therefore not known in as much detail as those of *Rellimia*. However, Scheckler and Banks (1971b) observe that they were probably smaller and less complex, but comparable morpho-

F

logically, dichotomizing at least once, and bearing pairs of sporangia pinnately.

Tetraxylopteris is known from Upper Devonian rocks of New York state (Beck, 1957; Bonamo and Banks, 1967). Fig. 38E illustrates the opposite and decussate mode of branching of the lateral branches, which were themselves arrange spirally on a main axis. In keeping with this external branching pattern, the primary xylem strand was cruciform. There were small groups of parenchyma, here and there within the arms of the primary xylem, but not sufficient to justify describing it as medullated. The secondary wood was compact and composed of tracheids with bordered pits on all the walls (i.e. tangential as well as radial).

The ultimate segments of sterile branches (not shown in Fig. 38) were once or twice bilobed and were not flattened at all. Fertile regions are illustrated in Figs. 38F and G, which have been re-drawn from reconstructions by Bonamo and Banks (1967). They consisted of main axes, bearing opposite and sub-opposite sporangial complexes which dichotomized twice. Each resulting branch was thrice pinnate, as shown in Fig. 38G, and the ultimate divisions bore sporangia which were oval and up to 5 m long. The general similarity of the fertile regions to those of *Rellimia* is striking – so much so that Bonamo and Banks remark that they cannot be distinguished when isolated from the rest of the plant.

Sphenoxylon is the name given to a single axis that was found in

---

Fig. 38   Aneurophytales

A, B, *Rellimia* (= *Protopteridium*): A, early reconstruction of branch system; B, ultimate fertile branchlets, as elucidated by *dégagement* (terminal sporangia omitted). C, D, *Aneurophyton*: C, early reconstruction of whole plant, based on trunks known as *Eospermatopteris* and branch systems thought, at the time, to have been fronds; D, frond-like branch system with minute forked 'leaves'. E–G, *Tetraxylopteris*: E, lateral branch system; F, portion of fertile branch system; G, ultimate branchlets of fertile branch system, with terminal sporangia (solid black). H, *Triloboxylon*, reconstruction showing spiral branching and leaves dichotomizing in one plane. I, J, *Proteokalon*: I, reconstruction, showing decussate branching (and occasional dichotomies, indicated by arrow); J, leaf, showing dichotomies in one plane (the degree of flattening is indicated by t.s. at various places indicated by arrows).

(A, C, D, after Kräusel and Weyland; B, Leclercq and Bonamo; E, H–J, Scheckler and Banks; F, G, Bonamo.)

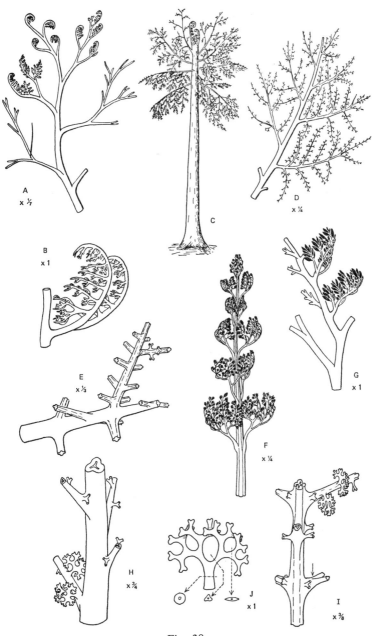

Fig. 38

marine strata of Upper Devonian age. Before becoming fossilized, it had become stripped of its bark, so that only some secondary wood and the primary xylem became petrified. Matten and Banks (1967) showed that it was identical with *Tetraxylopteris*, but suggest that the name *Sphenoxylon* should, nevertheless, be retained for specimens whose complete identity cannot be established.

Fossil remains of *Triloboxylon* have been found in Upper, as well as Middle, Devonian rocks of New York state (Scheckler and Banks, 1971a). Two species are known, one of which was formerly placed in the genus *Aneurophyton*. The primary vascular strand in the lateral branches was three-armed, with up to sixteen mesarch proto-xylem areas (and, in one of the two species, small areas of parenchyma) associated with departing lateral traces. Branching was spiral throughout and the ultimate branches bore appendages which forked four or more times in one plane (Fig. 38H).

The genus *Proteokalon*, also from Upper Devonian rocks of New York state, combined some of the features of *Tetraxylopteris* with some of those of *Triloboxylon* (Scheckler and Banks, 1971b). Thus, its lateral branch systems branched in an opposite and decussate fashion (like those of *Tetraxylopteris*, except that occasional dichotomies occurred – Fig. 38I), while its sterile appendages were like those of *Triloboxlyon*, in that they forked several times in one plane. However, the sterile appendages were much more leaf-like than those of *Triloboxylon*, in that their ultimate segments were flattened (Fig. 38J).

Unfortunately, no reproductive organs have yet been found belonging to either *Triloboxylon* or to *Proteokalon*.

PROTOPITYALES

Protopitys has been known for over a century as a stem genus from Lower Carboniferous deposits but only recently has anything been known of its reproductive organs. Trunks of *P. Buchiana* attained a diameter of at least 45 cm and must have grown to a considerable height. In the centre was an elliptical pith surrounded by a narrow zone of metaxylem. Lateral traces were given off from the opposite ends of the ellipse. They have always been described as 'leaf traces' but, in the light of what we now know of other progymnosperms, this must be questioned. However, whether they were leaf traces or branch traces, the tree must have had a strange appearance, likened by Scott (1923) to the Traveller's Tree of Madagascar (*Ravenala*).

The secondary wood of *Protopitys* was dense, like that of modern conifers, and had very small wood-rays, only one or two cells wide and frequently only one cell high. The tracheids had bordered pits, often uniseriate, restricted to the radial walls. Such wood was thought, for a long time, to be characteristic of gymnosperms (i.e. seed plants). It was therefore of the greatest interest that Walton (1957) demonstrated some slight degree of heterospory in *P. scotica*. He described a small stem, 7 cm long and 6 mm in diameter, bearing two lateral branch systems terminating in elongated sporangia 3 mm long. Most of the spores were 82 μ in diameter, but some sporangia contained larger ones, 147 μ across. However, the distinction was not clear cut, for some sporangia contained spores of an intermediate size, 98 μ. On the basis of these findings, Walton suggested that *Protopitys* should be regarded as a pteridophyte exhibiting the early stages of evolution of heterospory.

ARCHAEOPTERIDALES

Just as the trunks of *Protopitys* were originally believed to belong to gymnospermous plants, so also were those of *Callixylon*, of which some four species are known from the Upper Devonian and Lower Carboniferous rocks of North America and one from Russia. Trunks have been found up to 1·5 m in diameter and of an estimated overall height of at least 20 m. The bulk of the trunk was made up of secondary wood, with tracheids whose radial walls had bordered pits arranged in a peculiar manner. The pits were in groups separated by unpitted regions, and the pattern was the same on adjacent tracheids, so that in radial section (Fig. 39A) one sees horizontal bands of pits separated by unpitted bands. In most species, the wood-rays were only one or two cells wide (Fig. 39B), but in one species, *C. Newberryi*, the rays were as much as four cells wide. In the centre of the trunk there was a pith region, 1 cm or more across, with numerous mesarch strands of primary xylem near the periphery. From these 'circum-medullary strands', lateral traces (originally described as 'leaf-traces') had their origin. Except for the peculiar pitting of *Callixylon*, these trunks were so similar to those of modern conifers that they were almost universally believed to have been early members of the group, and it was widely believed that, when its leaves were found, they would be needle-shaped or strap-shaped.

It came as a great shock, therefore, when Beck (1960) demonstrated organic connection between branches identified as *Callixylon Zalesskyi* and large frond-like structures known as *Archaeopteris macilenta*. Some six species of *Archaeopteris* are known, all of which superfici-

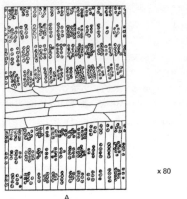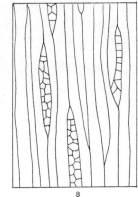

x 80

A   B

Fig. 39   Callixylon

A, B, radial longitudinal and tangential longitudinal sections,
respectively, of secondary wood of *C. Newberryi*. (Note, in A, the way
in which the pits on the radial walls of the tracheids are arranged in
groups.)

ally resemble fern fronds, some being at least a metre long, branching
bi-pinnately, with the pinnae all in one plane, e.g. *A. latifolia* (Fig.
40A). Terminology appropriate to fern fronds was used to describe
them, the ultimate segments being referred to as pinnules. In some
species, the sterile 'pinnules' were entire and fan-shaped, as in *A.
latifolia* (possibly synonymous with *A. hibernica* – Fig. 40E), while
in others they were deeply divided and laciniate, or fimbriate, as in
A. fissilis (Fig. 40F). The fertile 'pinnules' were always slender or
finely divided, and bore numerous elongated sporangia in two rows
on the adaxial side. Dehiscence was by means of a longitudinal slit.

At least three species of *Archaeopteris* are now known to have
been heterosporous. This was first demonstrated by Arnold (1939)
in *A. latifolia*. Figs. 40C and D illustrate the contents of a microsporangium and a megasporangium respectively. The former contained a hundred or more spores about 35 μ in diameter, while the
latter contained eight or sixteen about 300 μ in diameter. More
recently heterospory has been established in *A. Halliana* and *A.
macilenta*, where the megaspores are about seventeen times bigger
than the microspores (Phillips, Andrews and Gensel, 1972). There
can be no doubt, therefore, that these species of *Archaeopteris* were
at the level of a pteridophyte in their reproductive processes, for they
cannot have borne seeds in addition to megasporangia.

As long as it was thought that the sporangia were borne on fronds,

Archaeopteris (+*Callixylon*) was regarded as a most extraordinary kind of plant, fitting into none of the approved categories of plant organization. However, recent work (Carluccio, Hueber and Banks, 1966; Beck, 1971; Phillips, Andrews and Gensel, 1972) has shown that what were thought to be fronds were really branch systems and

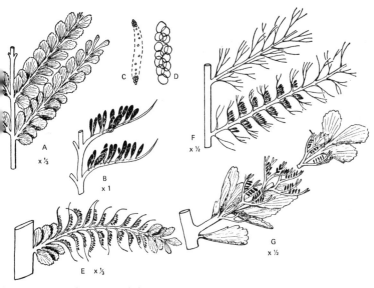

Fig. 40　Archaeopteridales

A–D, *Archaeopteris latifolia* (=*S. hibernica*?): A, part of frond-like lateral branch system; B, fertile leaves; C, D, contents of micro-sporangium and megasporangium, respectively. E, fertile branch of *A. hibernica*. F, portion of branch system of *A. fissilis*, with one sterile and one partially fertile branchlet. G, branchlet of *A. Halliana*, showing spiral arrangement of sterile and fertile leaves.

(A–D, after Andrews; E, Schimper; F, Andrews, Phillips and Radforth; G, Phillips, Andrews and Gensel.)

that the 'pinnules' were really small leaves. This was first suspected when it was shown that their vascular traces arose in spiral sequence from a radially symmetrical vascular system within the branches. It was then confirmed by *dégagement*, which showed that the ultimate appendages (i.e. the so-called 'pinnules') were indeed attached spirally, and could be traced down into the rock. The frond-like appearance (Fig. 40A) is explained as an artifact resulting from the splitting of the rock. Only those leaves which were in the plane of the fracture are visible, and they do indeed lie in one plane. The true three-dimensional arrangement of the leaves, both fertile and sterile, is illustrated for *A. Halliana* in Fig. 40G. In this species, the fertile leaves dichotomized up to four times and bore up to 20 sporangia along the adaxial face, except right at the tips. (In Fig. 40G the sporangia attached at the initial dichotomy have been omitted for clarity.)

In the light of this new interpretation, *Archaeopteris* and *Callixylon* together exhibit the type of organization which is found in the conifers and their immediate ancestors, viz. dense secondary wood, lateral branch systems which are often in one plane, and small leaves arranged spirally. It has been suggested that they might also have provided a starting point for the evolution of compound leaves, such as those of cycads. All that was needed was for the mode of attachment of the leaves to change and for the internal anatomy of the branches to become dorsiventral. However, such changes, although sounding simple, are in fact rather far-reaching and many morphologists would regard them as rather improbable.

For many years, *Callixylon* was classified along with two Lower Carboniferous stem genera, *Pitys* and *Archaeopitys*. Like *Callixylon*, *Pitys* grew to large dimensions, and some specimens must have been much more than 20 m tall. At least five species have been described, all of which have a relatively wide pith in the centre, up to 5 cm across. In some species, medullation was incomplete, in that there was a 'mixed pith' of tracheids and parenchyma. Even in relatively small stems, there were as many as 50 mesarch primary xylem strands in the peripheral regions. Lateral traces, described as 'leaf traces', arose as branches from these circum-medullary strands and subdivided into three on their way through the secondary wood and cortex. The secondary wood varied considerably from species to species in the size of the wood-rays for, whereas in *P. Withamii* they were only two or three cells wide, in *P. primaeva* and *P. rotunda* they were up to 15 cells wide.

What the leaves of *Pitys* were like is a matter of controversy. Gordon (1935) described them as short fleshy structures, only 5 cm long and 5 mm thick at the base. However, Beck (1960) points out that the leaf traces had a considerable quantity of secondary wood, which is more consistent with large fronds. At that time, it was believed that *Archaeopteris* was a frond genus, and Beck used the name Pityales to include *Pitys*, *Callixylon* and *Archaeopteris*. However, Long (1963) has cast some doubt on the status of *Pitys* as a pteridophyte. He draws attention to the frequent occurrence together of *Pitys primaeva* stems and petioles, known as *Lyginorachis papilio*, whose internal anatomy is similar to that of the 'leaf' of *Pitys Dayi* described by Gordon. Now, *Lyginorachis* petioles are part of large fern-like fronds, the fertile parts of which, known as *Tristichia*, bore seeds. Accordingly, Long regards *Pitys* as the trunk of a gymnosperm. In view of this suggestion, the name Pityales has been abandoned in this book in favour of the new one Archaeopteridales, and *Pitys* is included only as a very doubtful member. *Archaeopitys*, represented by the one species *A. Eastmanii*, was so similar to *Pitys* that some palaeobotanists believe that it should be included in that genus.

Clearly, the Progymnospermopsida are of great interest to the student of gymnosperm evolution. They are also of great interest to the student of pteridophytes, for they are the only group which appears to have had the potentiality for further evolution into the gymnosperms and, thence, into the flowering plants which dominate the world's vegetation today. Where the progymnosperms came from is not clear but, of the various groups of the Psilopsida that preceded them in the fossil record, the Trimerophytales provide the most likely origins, since both groups were characterized by fusiform sporangia (Chaloner, 1972).

8 General Conclusions

In a book of this limited size it is impossible to describe in detail all the fossil plants that are known. Accordingly several major groups of vascular plants, many minor groups and a large number of genera have had to be omitted. Thus, there has been no mention of the Noeggerathiales, nor of the Pseudoborniales, on the grounds that they occupy isolated positions in the classification and throw almost no light at all on the evolution of modern plants. For details of these strange plants the reader is referred to textbooks of palaeobotany (e.g. Hirmer, 1927; Zimmermann, 1930; Arnold, 1947; Andrews, 1961; Boureau, 1964, 1967, 1970).

The last two decades have witnessed a great many exciting discoveries by palaeobotanists that have thrown light, not only on the interrelationships of the various groups within the pteridophytes, but also on the way in which seed plants may have been derived from them. Unfortunately, however, there are still no fossils which clearly link pteridophytes, in the reverse direction, with their possible ancestors. Discussions still take place as to whether pteridophytes evolved directly from Algae or from Bryophyta, and as to whether, in either case, they had a monophyletic or a polyphyletic origin. Until more fossils are known from the Ordovician, Cambrian and even the Pre-Cambrian, there would seem to be little hope of agreement on these matters. There are some, indeed, who doubt whether missing links' will ever be found. In the meantime, relying on what we know with certainty to have existed, we must guess at what their ancestors might have been like.

Subjective processes of this kind have led to a number of theories of land-plant evolution, of which the Telome Theory has had the greatest number of adherents since it was first propounded by Zimmermann (1930). According to this theory, all vascular plants evolved from a very simple leafless ancestral type, like *Rhynia*, made up of sterile and fertile axes ('telomes'). In order to explain the wide diversity of organization found in later forms, a number of trends are supposed to have occurred, in varying degrees in the different taxonomic groups. These are represented diagrammatically in Fig. 41 (1–5) and are called respectively (1) planation, (2) overtopping, (3) syngenesis, (4) reduction, (5) recurving.

Starting from a system of equal dichotomies in planes successively at right angles (A), planation leads to a system of dichotomies in one plane (B). Overtopping is the result of unequal dichotomies, and tends to produce a main axis with lateral branches (C); the culmination of this process is a monopodial system. Syngenesis results from the coalescence of apical meristems. When they fuse to form a marginal meristem ('foliar syngenesis'), a lamina with veins develops (D) and the process is called 'webbing'. Zimmermann also visualizes a second type of syngenesis ('axial syngenesis') in which several branches become absorbed into a single stout axis with a complex stelar anatomy (not shown in Fig. 41). While these three trends are

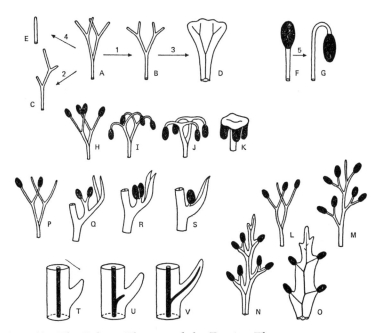

Fig. 41   The Telome Theory and the Enation Theory

The Telome Theory: 1, planation; 2, overtopping; 3, syngenesis; 4, reduction; 5, recurving. H–K, evolution in Sphenopsida; L–O, evolution in Pteropsida; P–S, evolution in Lycopsida.
The Enation Theory: T–V, evolution of microphylls in Lycopsida.

(A–S, based on Zimmermann; T–V, Bower.)

in the direction of progressive elaboration, the fourth is in the oppo-site direction, reduction, which is supposed to have brought about the evolution of simple unbranched microphylls from more complex structures. The fifth trend, re-curving, is found in several groups of plants, where the sporangiophore becomes reflexed and the sporan-gium inverted, as in anatropous ovules.

Figs. 41H–K illustrate the way in which the sporangiophore might have evolved in the Sphenopsida. Here, recurving and syngenesis are the chief trends, resulting in a peltate structure with reflexed sporangia like that of *Equisetum*. Intermediate stages were represented, among fossil members of the group by *Hyenia*, *Eviostachya* and *Protocalamo-stachys*. In the evolution of the leaf of the Sphenopsida, the chief trends are supposed to have been planation, followed by reduction. *Asterocalamites* provided an example of an intermediate stage in the process.

In the Pteropsida (Figs. 41L–O), planation, overtopping and webbing have combined to produce the sterile and fertile fronds of modern ferns. The fossil record provides abundant examples of in-termediate types of frond form (e.g. *Pseudosporochnus*, *Stauropteris*, *Botryopteris*). Little was known of the Progymnospermopsida at the time when the Telome Theory was first put forward, but they, too, provide excellent examples of intermediate stages in the evolution of leaves from stems, e.g. *Aneurophyton*, *Triloboxylon*, *Proteokalon* and *Archaeopteris*. Indeed, any plant in which there is no clear demerca-tion between leaf and stem can be said to lend support to the Telome Theory.

In the Lycopsida (Figs. 41P–S), the chief trend is supposed to have been reduction. The pentafid, trifid and bifid leaves and sporophylls of *Leclercqia*, *Colpodexylon* and *Protolepidodendron* respectively, might be quoted as examples of intermediate stages. However, if the microphylls of the Lycopsida had evolved by reduction in this way, it must be confessed that the fossil record gives no hint as to any earlier stages.

While the great appeal of the Telome Theory lies in its economy of hypotheses and in the way it allows the whole range of form of vascular plants to be seen in a single broad unified vista, neverthe-less, by some botanists it is regarded as a dangerous over-simplifica-tion, particularly in its application to the Lycopsida. Thus, the American palaeobotanist Andrews (1960) sums up his views in the following words: 'Zimmermann's scheme for the pteropsids, or at least some pteropsids, has much supporting evidence; his concept for the articulates may be valid, but we are only on the verge of understanding the origins of this group; his concept for the lycopsids is, so far as I am aware, purely hypothetical.'

Many morphologists prefer the Enation Theory of Bower (1935) which suggests that microphylls are not homologous in any way with megaphylls. According to this theory, microphylls started as bulges from the surface of the stem, and then evolved into longer and longer projections, at first without any vascular supply, then with a leaf trace that stopped short in the cortex of the stem and, finally, with a vascular bundle running the whole length of the organ. The microphyll, therefore, has evolved by a gradual process of enlargement, rather than by progressive reduction, and for this theory the fossil record does provide some support: *Psilophyton* represents the first stage in the process (Fig. 41T), *Asteroxylon* provides an example of the intermediate stage, where the leaf-trace stops short (Fig. 41U), while *Drepanophycus* represents a later stage with the leaf-trace entering the lateral appendage (Fig. 41V).

Before one can decide which of the two theories is the more appropriate, one needs to look very carefully at the time of appearance of the various fossil genera. It is not sufficient merely to observe whether it was Middle or Lower Devonian. One needs to know whether it was Gedinnian or Siegenian, etc. The starting point for the evolution of vascular plants, according to the Telome Theory, was a plant with naked forking axes and terminal appendages, and it has been customary to quote *Rhynia* and *Horneophyton* as examples. However, these were certainly not the ancestors of pteridophytes, for, as Leclercq (1954) emphasizes, they were among the last surviving examples of that particular growth form. It is hard to imagine a better example of the ideal 'telomic' plant than *Cooksonia*, which is also the earliest known vascular plant. From this early beginning, near the end of the Silurian, it is possible to envisage the evolution of the Trimerophytales, and from them in turn, the Pteropsida, Progymnospermopsida and, perhaps also, the Sphenopsida. It is equally possible to envisage the evolution of the Lycopsida from the Zosterophyllales, according to the Enation Theory, for both have lateral sporangia. What is difficult to imagine is how the Zosterophyllales could have evolved from a plant like *Cooksonia*, with terminal sporangia, even if there were sufficient time interval between the Upper Silurian and the Gedinnian. The possibility that the pteridophytes are, therefore, di-phyletic must be considered, but against this suggestion is the argument that Lycopsida have a number of anatomical features in common with other pteridophytes (e.g. cuticle with stomata, xylem and phloem) which are statistically unlikely to have come together more than once. Here, for the moment, the matter rests, there being some morphologists who believe in a monophyletic origin of the pteridophytes and some who believe in a diphyletic origin.

So far, these speculations as to the course of pteridophyte evolution have centred around the sporophyte, since it is this phase of the life-cycle that is represented in the fossil record. Even more speculative is the evolution of gametophytes, concerning which there are the two diametrically opposed schools of thought referred to, near the end of Chapter 3, as 'Antithetic' and 'Homologous'. Mention was there made of abnormal gametophytes of *Psilotum*, containing vascular tissues. The significance of this interesting discovery was somewhat diminished for a time, however, when it was shown that they were diploid; but in relation to discussions of antithesis and homology chromosome counts are, in a sense, 'red-herrings'. This is made apparent by the phemonena of apogamy and apospory, cases of which have been recorded many times in pteridophytes since they were first observed in 1874.

Apogamy is the development of a sporophyte directly from the gametophyte without the intermediate formation and fertilization of gametes. The resulting sporophyte, therefore, has the same haploid chromosome count as the gametophyte. By 1939 apogamy had been recorded among ferns, in *Pteris*, *Dryopteris*, *Pellaea*, and *Trichomanes*, where it is frequently preceded by the appearance of tracheids in the gametophyte (Steil, 1939). More recently (Bell, 1959) it has been recorded in *Thelypteris*, *Pteridium*, *Phyllitis* and several species of *Lycopodium*. In the case of *Phyllitis*, the haploid apogamous sporophyte was successfully reared until it produced sporangia; however, as would be expected since it contained only one set of chromosomes meiosis failed and no spores were produced (Manton, 1950).

Apospory is, in a sense, the reverse process, being the production of gametophytes directly from sporophytes without the intermediate formation of spores. Thus, when detached pieces of fern fronds are placed on an agar surface they frequently develop directly into gametophytes of normal shape and form. In such cases, the gametophyte has the same diploid chromosome count as the sporophyte. So numerous are the recorded instances of this phenomenon that Bell (1959) suggests that it must be general among ferns; yet the exact conditions under which it happens cannot yet be specified.

As to the causes of apogamy, several theories have been put forward, but the final word has certainly not been said on this fascinating subject. In many cases, ageing of the prothallus seems to be an important factor. Recent work in America (Whittier and Steeves, 1960) on *Osmunda*, *Adiantum* and *Pteridium* has, however, demonstrated that apogamy can be induced by growing the prothalli on an agar culture medium rich in glucose. Clearly, therefore, under these highly artificial circumstances, the external environment can be an important factor. That this might be so had been suspected for a long time, since otherwise it was difficult to understand why a diploid zygote developing inside a fertilized archegonium should give rise to a sporophyte, while a diploid cell developing by apospory should give rise to a gametophyte. Confirmation of the view that the internal environment of the archegonium exerts an important formative influence on the nature of the embryo has recently come from experiments in which young embryos of *Todea* were dissected from the archegonium and grown on an artificial medium (De Maggio and Wetmore, 1961). It was found that those removed before the first division of the zygote developed into flat thalloid structures, whereas those removed in later stages of development grew into normal sporophytes. Whether the environment is entirely reponsible, however, for the normal regular alternation of generations has been questioned. Bell (1959) suggests that there must be some internal factor at work and looks upon gametophyte and sporophyte as two levels of complexity, reflecting different states of the cytoplasm, which can be accounted for in terms of cell chemistry. This interesting hypothesis should stimulate further research into the causes of alternation of generations in living plants.

The present position, then, seems to be that there is no fundamental distinction between gametophytes and sporophytes, since they can be induced to change from one to the other in either direction. They are 'homologous', as far as can be judged from living plants, and one is led to speculate, therefore, that they were probably alike in form and structure in the earliest ancestors of land plants. Merker's suggestion, already mentioned in Chapter 2, that the horizontal axes of the Rhyniaceae were gametophytes, instead of sporophytic rhizomes, is of enhanced interest, therefore, because if confirmed it will provide the only kind of evidence which can really settle the controversy. As with most problems of macro-evolution, it is the palaeobotanist who has the key within his reach.

Bibliography

Agashe, S. N., 1964. *Phytomorph.*, **14**, 598–611. (Extraxylary tissues in *Calamites*)

Andrews, H. N., 1960. *Cold Spring Harb. Symp. quant. Biol.*, **24**, 217–34. (Early vascular plant evolution)

Andrews, H. N., 1961. *Studies in Paleobotany.* Wiley, New York and London.

Andrews, H. N., Kasper, A., and Mencher, E., 1968. *Bull. Torrey bot. Club*, **95**, 1–11. (*Psilophyton Forbesii*)

Andrews, H. N., and Mamay, S. H., 1951. *Bot. Gaz.*, **113**, 158–65. (*Bowmanites bifurcatus*)

Andrews, H. N., and Phillips, T. L., 1968. *J. Linn. Soc. (Bot.)*, **61**, 37–64. (*Rhacophyton ceratangium*)

Arber, A., 1950. *The Natural Philosophy of Plant Form.* Cambridge University Press, London.

Arnold, C. A., 1939. *Contr. Mus. Paleont. Univ. Mich.*, **5**, 217–314. (Heterospory in *Archaeopteris*)

Arnold, C. A., 1947. *An Introduction to Paleobotany.* McGraw-Hill, New York and London.

Arnold, C. A., 1960. *Contr. Mus. Paleont. Univ. Mich.*, **15**, 249–67. (Cambium in a Lepidodendrid)

Banks, H. P., 1944. *Am. J. Bot.*, **31**, 649–59. (*Colpodexylon*)

Banks, H. P., 1968. In *Evolution and Environment*, Drake, E. T., (Ed.). Yale University Press. (Early history of land plants)

Banks, H. P., Bonamo, P. M., and Grierson, J. D., 1972. *Rev. Palaeobot. Palynol.*, **14**, 19–40. (*Leclercqia*)

Banks, H. P., and Davis, M. R., 1969. *Am. J. Bot.*, **55**, 436–49. (*Crenaticaulis*)

Banks, H. P., and Grierson, J. D., 1968. *Palaeontographica*, **B, 123**, 113–20. (*Drepanophycus*)

Barber, H. N., 1957. *Proc. Linn. Soc. N. S. W.*, **82**, 201–8. (Polyploidy in Psilotales)

Baxter, R. W., 1963. *Am. J. Bot.*, **50**, 469–76. (*Calamocarpon*)

Baxter, R. W., 1967. *Kans. Univ. Sci. Bull.*, **47**, 1–23. (*Litostrobus*)

Baxter, R. W., and Leisman, G. A., 1967. *Am. J. Bot.*, **54**, 748–54. (*Calamostachys*)

Beck, C. B., 1957. *Am. J. Bot.*, **44**, 350–67. (*Tetraxylopteris*)

Beck, C. B., 1960. *Brittonia*, **12**, 351–68. (*Archaeopteris* and *Callixylon*)

Beck, C. B., 1971. *Am. J. Bot.*, **58**, 758–84. (*Archaeopteris*)

Bell, P. R., 1959. *J. Linn. Soc.* (*Bot.*), **56**, 188–203. (The pteridophyte life-cycle)

Bierhorst, D. W., 1954a. *Am. J. Bot.*, **41**, 274–81. (Gametangia and embryo of *Psilotum*)

Bierhorst, D. W., 1954b. *Am. J. Bot.*, **41**, 732–9. (Rhizome of *Psilotum*)

Bierhorst, D. W., 1956. *Phytomorph.*, **6**, 176–84. (Aerial shoots of Psilotales)

Bierhorst, D. W., 1958. *Am. J. Bot.*, **85**, 416–33, 534–37. (Xylem in *Equisetum*)

Bierhorst, D. W., 1959. *Am. J. Bot.*, **46**, 170–9. (Symmetry in *Equisetum*)

Bierhorst, D. W., 1960. *Phytomorph.*, **10**, 249–305. (Tracheary elements in vascular plants)

Bierhorst, D. W., 1966. *Am. J. Bot.*, **53**, 123–33. (Prothallus of *Actinostachys*)

Bierhorst, D. W., 1967. *Am. J. Bot.*, **54**, 538–49. (*Schizaea dichotoma*)

Bierhorst, D. W., 1968a. *Am. J. Bot.*, **55**, 87–108. (*Schizaea* and *Actinostachys*)

Bierhorst, D. W., 1968b. *Phytomorph.*, **18**, 232–68. (Stromatopteridaceae and Psilotaceae)

Bierhorst, D. W., 1969. *Am. J. Bot.*, **56**, 160–74. (*Stromatopteris*)

Bierhorst, D. W., 1971. *Morphology of Vascular Plants*. Macmillan, London and New York.

Bierhorst, D. W., 1973. In *The Phylogeny and Classification of the Ferns*, Jermy, A. C., Crabbe, J. A., and Thomas, B. A. (Eds.). Academic Press, London and New York. (Non-appendicular fronds in ferns)

Bonamo, P. M., and Banks, H. P., 1967. *Am. J. Bot.*, **54**, 755–68. (Fertile *Tetraxylopteris*)

Boureau, E. (Ed.), 1964, 1967, 1970. *Traité de Paléobotanique*, vols. 3, 2 and 4, respectively. Masson, Paris.

Bower, F. O., 1923, 1926, 1928. *The Ferns*, vols. 1–3. Cambridge University Press, London.

Bower, F. O., 1935. *Primitive Land Plants, also known as the Arche-*

goniatae. Macmillan, London. (Reprinted, 1959, Haffner, New York)

Browne, I. M. P., 1927. *Ann. Bot.,* **41**, 301–20. (Calamarian cones)

Bruchmann, H., 1897. *Untersuchungen über* Selaginella spinulosa. Perthes, Gotha.

Bruchmann, H., 1898. *Über die Prothallien und die Keimpflanzen mehrerer europäischer Lycopodien.* Perthes, Gotha.

Bruchmann, H., 1909a. *Flora,* **99**, 12–51. (Embryology of *Selaginella*)

Bruchmann, H., 1909b. *Flora,* **99**, 193–202. (Chemotaxis in *Lycopodium* sperms)

Bruchmann, H., 1912, 1913. *Flora,* **104**, 180–224, **105**, 337–46 (Embryology in *Selaginella*)

Campbell, E. O., 1936. *Trans. R. Soc. N. Z.,* **66**, 1–11. (*Histiopteris*)

Carluccio, L. M., Hueber, F. M., and Banks, H. P., 1966. *Am. J. Bot.,* **53**, 719–30. (*Archaeopteris*)

Chaloner, W. G., 1967. *Rev. Palaeobot. Palynol.,* **1**, 83–93. (Spores and early land plants)

Chaloner, W. G., 1972. *Rev. Palaeobot. Palynol.,* **14**, 49–61. (Devonian plants from Fair Isle)

Chaphekar, M., 1963. *Palaeontology,* **6**, 408–29. (*Protocalamites*)

Chaphekar, M., and Alvin, K. L., 1972. *Rev. Palaeobot. Palynol.,* **14**, 63–76. (Fertile *Metaclepsydropsis*)

Chiarugi, A., 1960. *Caryologia,* **13**, 27–150. (Table of chromosome numbers in pteridophytes)

Davie, J. H., 1951. *Am. J. Bot.,* **38**, 621–8. (Antheridial development in filicales)

Dawson, J. W., 1859. *Quart. J. geol. Soc. Lond.,* **15**, 477–88. (Devonian plants of Canada)

De Maggio, A. E., and Wetmore, R. H., 1961. *Am. J. Bot.,* **48**, 551–65. (Culture of excised fern embryos)

Duckett, J. G., 1970a. *New Phytol.,* **69**, 333–46. (Spore size in *Equisetum*)

Duckett, J. G., 1970b. *Bot. J. Linn. Soc.,* **63**, 327–52. (Sexual behaviour of *Equisetum*)

Duckett, J. G., 1972. *Bot. J. Linn. Soc.,* **65**, 87–108. (Sexual behaviour of *Equisetum*)

Duckett, J. G., 1973. *Bot. J. Linn. Soc.,* **66**, 1–22. (Gametophytes of *Equisetum*)

Eames, A. J., 1936. *Morphology of Vascular Plants (Lower Groups).* McGraw–Hill, New York and London.

Edwards, D., 1969. *Am. J. Bot.,* **56**, 201–10. (*Zosterophyllum*)

Edwards, D., 1970a. *Palaeontology,* **13**, 451–61. (*Cooksonia*)

Edwards, D., 1970b. *Phil. Trans. R. Soc.*, **B, 258**, 225–43. (*Gosslingia*)

Edwards, D., 1973. In *Atlas of Palaeobiogeography*, Hallam, A. (Ed.). Elsevier, Amsterdam. (Devonian floras)

Eggert, D. A., 1961. *Palaeontographica*, **B, 108**, 43–92. (Ontogeny of *Lepidodendron*)

Eggert, D. A., 1963. *Am. J. Bot.*, **50**, 379–87. (*Ankyropteris*)

Eggert, D. A., 1974. *Am. J. Bot.*, **61**, 405–13. (Branching sporangia in *Horneophyton*)

Eggert, D. A., and Gaunt, D. D., 1973. *Am. J. Bot.*, **60**, 755–70. (Secondary phloem in *Sphenopyhllum*)

Eggert, D. A., and Taylor, T. N., 1966. *Palaeontographica*, **B, 118**, 52–73. (*Tedelea = Ankyropteris*)

El-Saadawi, W. El-S., 1966. Ph.D. dissertation, University of Wales. (Branched sporangia in *Horneophyton*)

Engler, A., and Prantl, K., 1902. *Die natürlichen Pflanzenfamilien*. Engelmann, Leipzig.

Fabbri, F., 1963, 1965. *Caryologia*, **16**, 237–335; **18**, 675–731. (Supplements to Chiarugi, A., 1960, tables of chromosome numbers in pteridophytes)

Fairon, M., 1967. *Annls Soc. géol. Belg.*, **10**, 1–30. (*Asteroxylon elberfeldense*)

Farrar, D. R., and Wagner, W. H., 1968. *Bot. Gaz.*, **129**, 210–19. (Gametophytes of *Trichomanes*)

Ford, S. O., 1904. *Ann. Bot.*, **18**, 589–605. (*Psilotum*)

Frankenberg, J. M., and Eggert, D. A., 1969. *Palaeontographica*, **B, 128**, 1–47. (*Stigmaria*)

Freeberg, J. A., and Wetmore, R. H., 1957. *Phytomorph.*, **7**, 204–17. (*Lycopodium* prothalli *in vitro*)

Fry, W. L., 1954. *Am. J. Bot.*, **41**, 415–28. (*Paurodendron*)

Galtier, J., 1969. *C. r. hebd. Séanc. Acad. Sci., Paris.* **268**, 3025–8. (*Botryopteris antiqua*)

Galtier, J., 1971. *Naturalia monspel.*, **22**, 145–55. (Sporangia of *Botryopteris forensis*)

Gensel, P., Kasper, A., and Andrews, H. N., 1969. *Bull. Torrey bot. Club*, **96**, 265–76. (*Kaulangiophyton*)

Gibson, R. J. H., 1894. *Ann. Bot.*, **8**, 133–206. (Stem anatomy of *Selaginella*)

Gibson, R. J. H., 1896. *Ann. Bot.*, **10**, 77–88. (Ligules of *Selaginella*)

Gibson, R. J. H., 1897 *Ann. Bot.*, **11**, 123–55. (Leaves of *Selaginella*)

Gibson R. J. H., 1902. *Ann. Bot.*, **16**, 449-66. (Roots of *Selaginella*)

Goldring, W., 1924. *Bull. N. Y. St. Mus.*, **251**, 50–72. (*Eospermatopteris*)

Gordon, W. T., 1911. *Trans. R. Soc. Edinb.*, **48**, 163–90. (*Metaclepsydropsis*)

Gordon, W. T., 1935. *Trans. R. Soc. Edinb.*, **58**, 279–311. (*Pitys*)

Grierson, J. D., and Banks, H. P., 1963. *Palaeontogr. amer.*, **4**, 219–95. (Devonian lycopods)

Härtel, K., 1938. *Beitr. biol. Pfl.*, **25**, 125–68. (Stem apices of lycopods)

Harland, W. B., Smith, A. G., and Wilcock, B. (Eds.), 1964. *The phanerozoic time scale*. Geol. Soc., London.

Heard, A., 1927. *Quart. J. geol. Soc. Lond.*, **85**, 195–209. (*Gosslingia*)

Hickling, G., 1907. *Ann. Bot.*, **21**, 369–86. (*Palaeostachya vera*)

Hirmer, M., 1927. *Handbuch der Paläobotanik*, vol. 1, *Thallophyta, Bryophyta, Pteridophyta*. Oldenbourg, München and Berlin.

Holloway, J. E., 1917. *Trans. N. Z. Inst.*, **50**, 1–44. (*Tmesipteris*)

Holloway, J. E., 1921. *Trans. N. Z. Inst.*, **53**, 386–422. (*Tmesipteris*)

Holloway, J. E., 1939. *Ann. Bot.*, **3**, 313–36. (Prothalli of *Psilotum*)

Holttum, R. E., 1949. *Biol. Rev.*, **24**, 267–96. (Classification of ferns)

Holttum, R. E., 1954. *Flora of Malaya*, vol. 2, *Ferns of Malaya*. Government Printing Office, Singapore.

Holttum, R. E., 1971. *Blumea*, **19**, 17–52. (Thelypteridaceae)

Holttum, R. E., and Sen, U., 1961. *Phytomorph.*, **11**, 406–20. (Tree ferns)

Hopping, C. A., 1956. *Proc. R. Soc. Edinb.*, **66**, 10–28. (*Trimerophyton*)

Hoskins, J. H., and Abbott, M. L., 1956. *Am. J. Bot.*, **43**, 36–46. (*Selaginellites crassicinctus*)

Hoskins, J. H., and Cross, A. T., 1943. *Am. Midl. Nat.*, **30**, 113–63. (*Bowmanites = Sphenophyllostachys*)

Hueber, F. M., 1971. *Taxon*, **20**, 641–2. (*Sawdonia*, new name for *Psilophyton princeps*, var. *ornatum*)

Hueber, F. M., 1972. *Rev. Palaeobot. Palynol.*, **14**, 113–27. (*Rebuchia*)

Hueber, F. M., and Banks, H. P., 1967. *Taxon*, **16**, 81–5. (*Psilophyton* redefined)

Jermy, A. C., Crabbe, J. A., and Thomas, B. A. (Eds.), 1973. *The Phylogeny and Classification of the Ferns*. (Supplement no. 1 to *Bot. J. Linn. Soc.*, **67**). Academic Press, London.

Jones, C. E., 1905. *Trans. Linn. Soc. Lond.*, **7**, 15–36. (Stem anatomy of *Lycopodium*)

Joy, K. W., Willis, A. J., and Lacey, W. S., 1956. *Ann. Bot.*, **20**, 635–7. (Rapid fossil-peel technique)

Kasper, A. E., and Andrews, H. N., 1972. *Am. J. Bot.*, **59**, 897–911. (*Pertica*)

Kidston, R., and Gwynne-Vaughan, D. T., 1907, 1908, 1909, 1910 and 1914. *Trans. R. Soc. Edinb.*, **45**, 759–80; **46**, 213–32; **46**, 651–67; **47**, 455–77; **50**, 469–80. (Fossil Osmundales)

Kidston, R., and Lang, W. H., 1917. *Trans. R. Soc. Edinb.*, **51**, 761–84. (*Rhynia Gwynne-Vaughani*)

Kidston, R., and Lang, W. H., 1920a. *Trans. R. Soc. Edinb.*, **52**, 603–27. (*Rhynia major* and *Horneophyton*)

Kidston, R., and Lang, W. H., 1920b. *Trans. R. Soc. Edinb.*, **52**, 643–80. (*Asteroxylon Mackiei*)

Kidston, R., and Lang, W. H., 1921a. *Trans. R. Soc. Edinb.*, **52**, 831–54. (Reconstructions of *Rhynia*, *Horneophyton* and *Asteroxylon*)

Kidston, R., and Lang, W. H., 1921b. *Trans. R. Soc. Edinb.*, **52**, 855–902. (Fungi in Rhynic plants)

Kräusel, R., and Weyland, H., 1923. *Senckenbergiana*, **5**, 154–84. (*Aneurophyton*)

Kräusel, R., and Weyland, H., 1926. *Abh. senckenb. naturforsch. Ges.*, **40**, 115–55. (*Asteroxylon, Aneurophyton, Hyenia, Cladoxylon, Calamophyton*)

Kräusel, R., and Weyland, H., 1929. *Abh. senckenb. naturforsch. Ges.*, **41**, 315–60. (*Aneurophyton*)

Kräusel, R., and Weyland, H., 1932. *Senckenbergiana*, **14**, 391–403. (*Protolepidodendron, Hyenia, Aneurophyton*)

Kräusel, R., and Weyland, H., 1933. *Palaeontographica*, **B**, **78**, 1–44. (*Protopteridium, Pseudosporochnus, Drepanophycus*)

Kräusel, R., and Weyland, H., 1935. *Palaeontographica*, **B**, **80**, 171–90. (Lower Devonian plants)

Lang, W. H., and Cookson, I. C., 1935. *Phil. Trans. R. Soc.*, **B**, **224**, 421–49. (*Baragwanathia, Yarravia*)

Leclercq, S., 1940. *Mém. Acad. r. Belg. Cl. Sci.*, **12**, 1–65. (*Hyenia*)

Leclercq, S., 1951. *Ann. Soc. geol. Belg.*, **9**, 1–62. (*Rhacophyton*)

Leclercq, S., 1954. *Svensk bot. Tidskr.*, **48**, 301–15. (Psilophytales)

Leclercq, S., 1957. *Mém. Acad. r. Belg. Cl. Sci.*, **14**, 1–40. (*Eviostachya*)

Leclercq, S., and Andrews, H. N., 1960. *Ann. Mo. bot. Gdn.*, **47**, 1–23. (*Calamophyton*)

Leclercq, S., and Banks, H. P., 1962. *Palaeontographica*, **B**, **110**, 1–34. (*Pseudosporochnus*)

Leclercq, S., and Bonamo, P. M., 1971. *Palaeontographica*, **B**, **136**, 83–114. (Fertile regions of *Rellimia = Milleria = Protopteridium*)

Leclercq, S., and Bonamo, P. M., 1973. *Taxon*, **22**, 435–37. (*Rellimia*, a new name for *Protopteridium*)

Leclercq, S., and Lele, K. M., 1968. *Palaeontographica*, **B, 123,** 97–112. (*Pseudosporochnus*)

Leclercq, S. and Schweitzer, H-J., 1965. *Bull. Acad. r. Belg. Cl. Sci.*, **51,** 1395–1403. (*Calamophyton*)

Leisman, G. A., and Bucher, J. L., 1971a. *Bull. Torrey bot. Club.*, **98,** 140–4. (*Palaeostachya*)

Leisman, G. A., and Bucher, J. L., 1971b. *J. Palaeontol.*, **45,** 494–501. (*Calamocarpon*)

Lemoigne, Y., 1968. *C. r. hebd. Séanc. Acad. Sci. Paris*, **266,** 1655–7. (Archegonia in *Rhynia*?)

Lemoigne, Y., 1970. *Bull. Soc. bot. Fr.,* **117,** 307–20. (Gametophytes of *Rhynia*?)

Long, A. G., 1963. *Trans. R. Soc. Edinb.*, **65,** 211–24. (Fronds associated with *Pitys*)

Lyon, A. G., 1964. *Nature*, **203,** 1082–3. (Fertile *Asteroxylon*)

McLean, R. C., and Ivimey-Cook, W. R., 1951. *Textbook of Theoretical Botany*, vol. 1. Longmans, London.

Mamay, S. H., 1950. *Ann. Mo. bot. Gdn.*, **37,** 409–76. (Carboniferous fern sori)

Mamay, S. H., 1954. *Ann. Bot.*, **18,** 229–40. (*Litostrobus*)

Manton, I., 1942. *Ann. Bot.*, **6,** 283–92. (Cytology of vascularized prothalli of *Psilotum*)

Manton, I., 1950. *Problems of cytology and evolution in the pteridophytes*. Cambridge.

Matten, L. C., and Banks, H. P., 1967. *Bull. Torrey bot. Club*, **94,** 321–33. (*Sphenoxylon* and *Tetraxylopteris*)

Melchior, H., and Werdermann, E., 1954. *Engler's Syllabus der Pflanzenfamilien*. Borntraeger, Berlin.

Merker, H., 1958, 1959. *Bot. Notiser*, **111,** 6C8–18; **112,** 441–52. (Gametophytes of Rhyniaceae?)

Morgan, J., 1959. *Illinois biolo. Monogr.* **no. 27.** (*Psaronius*)

Pant, D. D., 1962. In *Proc. Summer Sch. Bot. Darjeeling*, Maheshwari, P., Johri, B. M., and Vasil, I. K. (Eds.), 276–301. (Gametophytes of *Rhynia*?)

Phillips, T. L., Andrews, H. N., and Gensel, P. G., 1972. *Palaeontographica*, **B, 139,** 47–71. (Two heterosporous species of *Archaeopteris*)

Phillips, T. L., and Leisman, G. A., 1966. *Am. J. Bot.*, **53,** 1086–100. (*Paurodendron*)

Rauh, W., and Falk, H., 1959. Sber. heidelb. Akad. Wiss. (**1959**), 1–160. (*Stylites*)

Sahni, B., 1923. *J. Ind. Bot. Soc.*, **3,** 185–91. (Teratology of

Psilotales)

Sahni, B., 1928. *Phil. Trans. R. Soc.*, **B, 217**, 1–37. (*Austroclepsis*)

Satterthwait, D. F., and Schopf, J. W., 1972. *Am. J. Bot.*, **59**, 373–6. (Phloem of *Rhynia*)

Scheckler, S. E., and Banks, H. P., 1971a. *Am. J. Bot.*, **58**, 737–51. (*Triloboxylon* and *Tetraxylopteris*)

Scheckler, S. E., and Banks, H. P., 1971b. *Am. J. Bot.*, **58**, 874–84. (*Proteokalon*)

Schlanker, C. M., and Leisman, G. A., 1969. *Am. J. Bot.*, **130**, 35–41. (*Selaginella fraipontii*)

Schweitzer, H-J., 1968. *Palaeontographica*, **B, 123**, 43–75. (*Protocephalopteris*)

Schweitzer, H-J., 1972. *Palaeontographica*, **B, 137**, 154–75. (*Hyenia*)

Schweitzer, H-J., 1973. *Palaeontographica*, **B, 140**, 117–50. (*Calamophyton*)

Scott, D. H., 1920, 1923. *Studies in Fossil Botany*, vol. 1, *Pteridophyta*; vol. 2, *Spermophyta*. Black, London.

Sporne, K. R., 1949. *New Phytol.*, **48**, 259–76. (Advancement index in assessing phylogenetic status)

Sporne, K. R., 1956. *Biol. Rev.*, **31**, 1–29. (Circular classification)

Sporne, K. R., 1964. *Nature*, **201**, 1345–6. (Self-fertility in *Equisetum*)

Steil, W. N., 1939. *Bot. Rev.*, **5**, 433–53. (Apogamy and apospory in pteridophytes)

Stidd, B. M., 1971. *Palaeontographica*, **B, 134**, 87–123. (Fronds of *Psaronius*)

Stokey, A. G., 1951. *Phytomorph.*, **1**, 39–58. (Phylogeny of fern prothalli)

Stone, I. G., 1962. *Aust. J. Bot.*, **10**, 76–92. (Antheridium ontogeny in ferns)

Surange, K. R., 1952. *Palaeobotanist*, **1**, 420–34. (*Botryopteris antiqua*)

Taylor, T. N., 1967. *Am. J. Bot.*, **54**, 298–305. (*Calamostachys*)

Townrow, J. A., 1968. *J. Linn. Soc. Bot.*, **61**, 13–23. (A Permian *Selaginella*)

Treub, M., 1884. *Ann. Jard. bot. Buitenz.*, **4**, 107–38. (Prothallus of *Lycopodium cernuum*)

Treub, M., 1890. *Ann. Jard. bot. Buitenz.*, **8**, 1–37. (Embryo of *Lycopodium cernuum*)

Troop, J. E., and Mickel, J. T., 1968. *Am. Fern J.*, **58**, 64–70. (Petiolar shoots in ferns)

Tryon, A. F., 1964. *Am. J. Bot.*, **51**, 939–42. (Heterospory in *Platyzoma*)

Walton, J., 1949. *Trans. R. Soc. Edinb.*, **61**, 729–36. (*Protocalamo-stachys*)

Walton, J., 1957. *Trans. R. Soc. Edinb.*, **63**, 333–40. (*Protopitys*)

Walton, J., 1964. *Phytomorph.*, **14**, 155–60. (*Zosterophyllum*)

Wand, A., 1914. *Flora*, **106**, 237–63. (Apical meristems in *Selaginella*)

Ward, M., 1954. *Phytomorph.*, **4**, 1–17. (Embryology of *Phlebodium*)

Wardlaw, C. W., 1955. *Embryogenesis in Plants*. Methuen, London; Wiley, New York.

West, R. G., 1953. *New Phytol.*, **52**, 267–72. (*Azolla* in Interglacial deposits)

Whittier, D. P., and Steeves, T. A., 1960. *Can. J. Bot.*, **38**, 925–30. (Glucose causing apogamy in ferns and swelling of gametophyte)

Williams, S., 1933. *Trans. R. Soc. Edinb.*, **57**, 711–37. (Regeneration in *Lycopodium selago*)

Williamson, W. C., and Scott, D. H., 1894. *Phil. Trans. R. Soc.*, **B, 185**, 863–959. (*Palaeostachya vera*)

Zimmermann, W., 1930. *Phylogenie der Pflanzen*. Fischer, Jena.

Index

Page numbers in *italic* refer to illustrations